Failed Historical Scientific Instruments

Scientific Instruments and Collections

STUDIES PUBLISHED UNDER THE AUSPICES OF THE SCIENTIFIC INSTRUMENT COMMISSION

VOLUME 10

The titles published in this series are listed at *brill.com/sico*

Failed Historical Scientific Instruments

Edited by

Sara J. Schechner
Sofia Talas

BRILL

LEIDEN | BOSTON

Cover illustration: Victorian trade card, likely American. Schechner Collection.

Library of Congress Cataloging-in-Publication Data

Names: Schechner, Sara, 1957– editor. | Talas, Sofia, editor.
Title: Failed historical scientific instruments / edited by Sara J. Schechner, Sofia Talas.
Description: Leiden ; Boston : Brill, [2024] | Series: Scientific instruments and collections, 1877-7066 ; volume 10 | Includes bibliographical references and index. | Summary: "Can a scientific instrument be regarded as a failure? Why and how? By shedding light on the complexity of these questions, the volume marks a step forward in the way historical scientific instruments can be analysed and displayed. The essays show how diverse failures can be, and how the assessment of scientific devices may change over time – some surprisingly becoming more successful. In addition to studies of how technical features led to failure, the authors examine the roles played by social bias and behaviour, commercial and economic circumstances, and political factors"—Provided by publisher.
Identifiers: LCCN 2024019620 (print) | LCCN 2024019621 (ebook) | ISBN 9789004689091 (hardback) | ISBN 9789004689107 (ebook)
Subjects: LCSH: Scientific apparatus and instruments—History. | Scientific apparatus and instruments—Design and construction—History. | Errors, Scientific—History.
Classification: LCC Q185 .F35 2024 (print) | LCC Q185 (ebook) | DDC 502.8/4—dc23/eng/20240509
LC record available at https://lccn.loc.gov/2024019620
LC ebook record available at https://lccn.loc.gov/2024019621

Typeface for the Latin, Greek, and Cyrillic scripts: "Brill". See and download: brill.com/brill-typeface.

ISSN 1877-7066
ISBN 978-90-04-68909-1 (hardback)
ISBN 978-90-04-68910-7 (e-book)
DOI 10.1163/9789004689107

This book is printed on acid-free paper and produced in a sustainable manner.

Contents

List of Figures and Tables VII
Notes on Contributors XII

Introduction: Failed Historical Scientific Instruments 1
Sara J. Schechner and Sofia Talas

1 An Archive of Failed Inventions: the Museum of the Admiralty Compass Observatory 6
Richard Dunn

2 The Diffusion Hygrometer: a Commercial Failure 21
Jean-François Loude

3 "Extremely Poor": an Equatorial Mount That Failed 38
Johan Kärnfelt

4 A Telescope's Lack of Fortune: the Newall Telescope 57
Panagiotis Lazos and George N. Vlahakis

5 Not All Bad: Rehabilitating Phrenology in Museums 77
Tacye Phillipson

6 Finding Way at Sea – Alternatives to the Log-Line and Sand Glass. The British Experience in the Hanoverian Era: Conservative Navigational Practice or Technological Failure? 84
D. J. Bryden

7 The Transmission X-Ray Microscope: Super Microscope for Biology or Yet Another Niche Instrument? 132
Andreas Junk

8 From Complex to Simple: the Example of *Qibla*-Indicators 148
Razieh S. Mousavi

9 Pedro Nunes, Tycho Brahe, and Orazio Borgondio: Two Failed Schemes to Read Angular Scales? 167
Luís Tirapicos

10 Thomas Young's Eriometer: Wrong Time, Wrong Inventor, Wrong Users? 178
Michelle Mercier

11 Black and White Images in a Grey Area of Failure: the Photo-Theodolite at the Royal Geographical Society in the 1920s 193
Jane Wess

12 A Failure, an Icon or Beyond: the Cloud Chamber of the University of Padua 209
Sofia Talas and Giulio Peruzzi

13 Failures, an Instrument Maker's Perspective 226
Nicolàs de Hilster

General Index 237

Figures and Tables

Figures

0.1 This young lady finds her binoculars very unsatisfactory and might blame their optical quality, but in context, the causes of their failure include user errors and environmental factors. Victorian trade card, likely American 5
1.1 Patent drawing of the Jennings compass, showing the space for the treated iron filings, from Henry Constantine Jennings, *Specification of Henry Constantine Jennings. Mariners' Compass.* Eyre and Spottiswoode, London, 1857 15
1.2 Patent drawing of the card of the Jennings compass, showing the iron "shields", from Henry Constantine Jennings, *Specification of Henry Constantine Jennings. Mariners' Compass.* Eyre and Spottiswoode, London, 1857 15
1.3 Magnetic compass inscribed "insulating Compass No 6" and "Jennings & Cos. Patent", London, about 1818 18
1.4 Underside of the card of the Jennings compass with two iron "shields", London, about 1818 18
2.1 Example of the graphical determination of the relative humidity *RH* from the ambient and dew point temperature measurements, using the Tetens equation 26
2.2 Heinrich Greinacher's double diffusion hygrometer. *Schweizerische Bauzeitung* (1946), fig. 2 28
2.3 Schematic drawing of the double diffusion hydrometer. Adapted from Greinacher, *Helvetica Physica Acta* (1944), fig. 2 29
2.4 Greinacher's simplified diffusion hygrometer. *Chimia* (1953), fig. 3 30
2.5 Example of the graphical determination of the relative humidity from the manometer readout and the temperature. Adapted from Greinacher, *Schweizer Archive für angewandte Wissenschaft und Technik* (1949), fig. 2 31
2.6 Commercial diffusion hygrometers: (a) TESA and (b) Rüeger 32
2.7 Internal view of the TESA hygrometer 34
2.8 Hair hygrometer "Hygrofix Haenni" 36
3.1 The Herschel telescope of the Royal Swedish Academy of Sciences (RSAS), on its present mount (RSAS) 40
3.2 The 16 cm speculum mirror made by William Herschel and delivered to the Royal Swedish Academy of Sciences in 1790 (RSAS) 40
3.3 The Herschel telescope in its original mount, illustrated in a drawing by Jacob Johan Tavaststjerna, 1813 (RSAS) 45

3.4 Edward Nairne's 1771 portable equatorial telescope, as depicted in Edward Nairne, "XIV. Description and Use of a New Constructed Equatorial Telescope or Portable Observatory, Made by Mr Edward Nairne", *Philosophical Transactions of the Royal Society* 61 (1771), pp. 107–113 47

3.5 Telescope built by William Herschel for William Watson at Bath in 1783–1785 51

3.6 The Herschel telescope on exhibition in Stockholm in *Heritage from Newton and Linnaeus* in 1962 (RSAS) 54

4.1 The Newall telescope at Ferndene without its dome in 1877 62

4.2 Robert Newall on the observing platform mentioned by Henry Holiday, operates his telescope 66

4.3 Maurolycus crater on the Moon, sketched by Henry Holiday using the Newall telescope, 1870 67

4.4 Eratosthenes crater on the Moon, sketched by Henry Holiday using the Newall telescope, 1871 68

4.5 The dismantling of the dome for the Newall telescope at Cambridge in 1957 73

4.6 The dome of the Newall telescope at Penteli Observatory 74

4.7 The Newall telescope today 76

5.1 Cast plaster phrenological head with numbered sections cast into the head, made by Francisco Mitchelelle of Pathhead, Scotland, mid-nineteenth century 79

5.2 Set of silver phrenological callipers, presented to George Combe by the pupils of Mr Williams' Secular School, made by Michael Crichton, Edinburgh, 1852 80

5.3 Phrenology of the crow, from Joseph Vimont's *Traité de phrénologie humaine et comparée*, 1831 82

6.1 Charles Grant's perpetual log, or "Hydroscope", 1806 94

6.2 Richard Hall Gower's improved patent log, 1807, from Gower's *Supplement to the Practical Seamanship*, plate 4 94

6.3 James Hookey's fish-shaped log 101

6.4 Francis Higginson's spring-loaded log, circa 1828 101

6.5 R. R. Baines's perpetual log or "sea perambulator", 1816, from the *Repertory of Arts, Manufactures, and Agriculture*, 2nd series, 37, 1820 102

6.6 The second form of the Gould log, from the *Supplement to the Encyclopaedia or a Dictionary of Arts, Sciences and Miscellaneous Literature*, Philadelphia, 1803, vol. 3, appendix, plate 1 111

6.7 The Massey log, 1802 patent design 112

7.1 A zone plate and its structure of concentric rings 134
7.2 The modern setup of the TXM. Image adapted from Peter Fischer *et al.*, "Soft X-Ray Microscopy of Nanomagnetism", *Materials Today* 9, nos. 1–2 (2006), pp. 26–33 136
7.3 Experimental proof that zone plates could be used as imaging optics. The image shows the Fe XVII radiation from the Sun, recorded during ESRO Mission CR16 on 11 March 1971 139
7.4 The first successful recording with the TMX prototype shows a metallic copper grating, from B. Niemann, D. Rudolph, and G. Schmahl, "Soft x-ray Imaging Zone Plates with Large Zone Numbers for Microscopic and Spectroscopic Applications", *Optics Communications* 12, no. 2 (1974), pp. 160–163 140
7.5 The series of deteriorating red cells shows how the superficial and intracellular structures are reorganised after a malaria infection. From Cathleen Magowan *et al.*, "Intracellular Structures of Normal and Aberrant *Plasmodium Falciparum* Malaria Parasites Imaged by Soft X-Ray Microscopy," *Proceedings of the National Academy of Sciences* 94, no. 12 (June 1997), pp. 6222–6227, fig. 2 144
7.6 Tomographic "slice" of a plant cell (*left*), which together with other slices was computed into the cell's tomogram (*right*). Reprinted from D. Weiß *et al.*, "Computed Tomography of Cryogenic Biological Specimens Based on X-Ray Microscopic Images", *Ultramicroscopy* 84, nos. 3–4 (2000), pp. 185–197 145
7.7 An idea for a TXM icon: the alignment of fluorescence microscopy (*left*) with the red markers for the orientation and the TXM image (*right*) with the detailed information about the cells content and arrangement. Reprinted from Gerd Schneider *et al.*, "Cryo X-Ray Microscope with Flat Sample Geometry for Correlative Fluorescence and Nanoscale Tomographic Imaging", *Journal of Structural Biology* 177, no. 2 (2012), pp. 212–223 146
8.1 Finding *qibla* as a problem of spherical astronomy 151
8.2 The upper-right quarter of the back of this astrolabe is a quadrant for finding the direction of Mecca from some desired cities 154
8.3 A rough schema of *Maṭlaʿ al-anwār* devised by Muḥammad Bāqir Yazdī (a). Alongside numerous applications, the horizontal plate (b) offers three different methods for finding *qibla* of which the "method of *zījes*" is illustrated here 158
8.4 A Persian, rectangular *qibla*-indicator, now with empty housings for a hinged compass (*on top*) and a gnomon (*at the centre*). Unsigned and undated, possibly eighteenth century 160
8.5 *Ṣafḥa-yi jahān-numā* ("Plate of the World"): a Mecca-centred *qibla*-indicator accompanying a horizontal pin-gnomon sundial. Persian lines of instruction and a gazetteer of towns appear on the back of the instrument 162

8.6 A late *qibla*-indicator fitted with a compass and a horizontal pin-gnomon sundial. A gazetteer of numerous *qibla* arguments is engraved on its back, and on the inner and outer sides of its lid 163
9.1 Pedro Nunes's book *De Crepusculis* (1542), fol. 18r, where the author presents a diagram of the nonius scale 169
9.2 Engraving of the "medium azimuthal brass quadrant" with the nonius, in the second edition of Tycho Brahe's *Astronomiæ instauratæ mechanica*, Nuremberg, 1602 171
9.3 Azimuthal quadrant with a nonius scale, attributed to the English maker James Kynvyn, and produced for Robert Dudley, *c.*1595, as depicted in Dudley's *Dell'arcano del mare*, Florence, 1647 173
9.4 First page of Orazio Borgondio's paper describing his invention of "a new construction of an astronomical quadrant in which we have easily and with certainty the arc second", from *Mémoires pour l'histoire des sciences et des beaux arts* (April 1720), pp. 764–766 175
10.1 Thomas Young's eriometer. Part of a letter from Young to Sir Joseph Banks dated 10 September 1810 180
10.2 A schematic sketch of the experimental setup 181
10.3 A diffraction phenomenon created by wool fibres examined with a replica of an eriometer 184
10.4 Replica of Young's eriometer for Banks, 1810, built by Phil Hanisch 187
10.5 Experimental measurement of the diameter of small fibres by Paul Hanisch using his replica of Young's 1810 eriometer 188
11.1 The Stereoautograph of Von Orel and Zeiss, 1922, from Arthur R. Hinks, "Stereographic Survey: The Stereoautograph", *Geographical Journal* 59, no. 4 (April 1922), between pp. 276 and 277 199
11.2 Photograph of the Wild photo-theodolite acquired by the Royal Geographical Society in 1926, from D. A. Hutchison and Arthur Hinks, "The Photo-theodolite of Mr H. Wild", Geographical Journal 67, no. 3 (March 1926), between pp. 236 and 237 200
11.3 Hope Islands, mangrove and shingle cay, photograph by M. A. Spender, 1928 205
11.4 Kodak's shopfront, George Street, Sydney, 1932–1934 207
12.1 The newly built Institute of Physics in Padua, 1937 211
12.2 The Cockroft-Walton generator set up at Padua Institute of Physics, 1938 211
12.3 The cloud chamber kept at the Giovanni Pollen Museum in Padua 213
12.4 The cloud chamber described in 1953 by Cresti, Loria, and Zago, as depicted in Marcello Cresti, Arturo Loria, and Guido Zago, "Camera di Wilson in campo magnetico", *Il Nuovo Cimento* 10 (1953), pp. 843–850: 844 219

12.5 The cosmic-ray laboratory set up in 1950 by Padua Institute of Physics at Pian di Fedaia, close to the Marmolada, at an altitude of 2000 m in the Dolomites 220
12.6 Rossi's 7-ton electromagnet in use at the Fedaia cosmic-ray laboratory, in the 1950s 221
12.7 Marcello Cresti (*on the left*), and Guido Zago (*on the right*) with two tourists, at the entrance of the Fedaia cosmic-ray laboratory 222
13.1 Jennings insulated compass 226
13.2 Flow diagram of the creative process 227
13.3 The mariner's astrolabe as a function of circles 229
13.4 Optical tooling target from the author's workshop 230
13.5 The optical tooling target in action 232
13.6 Top view of the field cap of the Hubble Space Telescope mirror testing interferometer. NASA, November 1990, from Lew Allen *et al.*, *The Hubble Space Telescope: Optical Systems Failure Report*, NASA, November 1990, fig. 7-5 233
13.7 Repairing the Hubble Space Telescope. NASA, 5 December 1993, STS061-87-062, from https://spaceflight.nasa.gov/gallery/images/shuttle/sts-61/html/sts061-87-062.html (site discontinued) 234

Tables

1.1 The first fifty entries of "A List of Compasses, Instruments &c. at the Compass Observatory", about 1850 8
6.1 British log proposals in Hanoverian era, organised by date. All publications printed in London unless otherwise indicated 126

Notes on Contributors

David Bryden
held curatorial posts in Edinburgh, Cambridge, and London. His Cambridge PhD was awarded on evaluation of publications on early scientific instruments and the British instrument trade. In retirement he continues to research and publish. A monograph – *Innovation in the Design of Scientific Instruments in the Georgian Era: The Role of the Society of Arts*, William Shipley Research Group, Research Paper 3, 2019 – grew out of seeking comparative material to complement a long-standing and continuing investigation into Georgian patents for scientific instruments.

Richard Dunn
has been Keeper of Technologies and Engineering at the Science Museum, London, since 2019. Previously, he was Senior Curator for the History of Science at Royal Museums Greenwich. His publications include *The Telescope: A Short History* (2009), *Finding Longitude* (2014, with Rebekah Higgitt), and *Navigational Instruments* (2016).

Nicolàs de Hilster
is a researcher of instruments and methods for navigation, hydrography, and geodesy. His PhD thesis covers the history, construction, development, use, and accuracy of early modern instruments for celestial navigation, the subject in which he graduated at the Vrije Universiteit, Amsterdam in 2018. In addition, he is an amateur astronomer, collector of geodetic instruments, and is the representative Overseas Committee Member of the Scientific Instrument Society.

Andreas Junk
is Senior Lecturer for Physics and the History of Science at the Europa-Universität Flensburg. His research interests encompass subjects from the eighteenth to the twentieth centuries. Among these are the Feynman myth, the standard history of nanotechnology, and the history of optical instruments. His current research project deals with the early research on the physiology of human vision, and the subsequent development of stereoscopic instruments.

Johan Kärnfelt
is Associate Professor (docent) in History of Ideas and Science, working at the Department of Literature, History of Ideas, and Religion, at Gothenburg

University, Sweden. His main research interest is the Swedish history of science, in particular, history of astronomy. His focus has been mainly on the nineteenth and twentieth centuries, and amongst other things, he has studied popular science, amateur astronomy, and time standardisation. More recently, he has worked on the materialities of science, especially scientific instruments.

Panagiotis Lazos
is a physics teacher in secondary education and head of the Fourth Laboratory Centre of National Sciences of Athens. He is currently a PhD student in the Faculty of Primary Education of the National and Kapodistrian University of Athens. He is a member of several academic societies. His subjects of research are history of scientific instruments, history of science education, and history of the sciences in southeastern Europe (eighteenth to twentieth centuries).

Jean-François Loude
obtained his PhD in physics from the École Polytechnique Federal (EPFL) in 1969. At the University of Lausanne (UNIL), he collaborated on experiments at SIN and CERN, developing particle detectors and associated electronics. An associate professor from 1981, he retired in 2002 and returned to EPFL, where he started a new career centred on the history of physics instruments. He made inventories of the old instruments kept at UNIL and exhibited the most interesting examples in a small museum.

Michelle Mercier
studied mathematics and physics as an undergraduate. During her study she dealt with the history of physics, including Kunst's Tube (1868) and the thermoscope by Rumford (1804). She is now a postgraduate student at the Europa-Universität Flensburg in the institute of physics, its didactics, and its history. In her PhD project, she is working on Young's double-slit experiment (1807), employing the replication method.

Razieh S. Mousavi
is a research scholar at the Max Planck Institute for the History of Science in Berlin. She obtained her PhD from the Humboldt University of Berlin in October 2023. She specialises in the contextualised history of sciences in Islamic societies. Her work delves deeply into the realm of mathematical practices, the integration of astronomical knowledge into various realms of pre-modern scholarship, and the interplay between science and religion within Muslim societies.

Giulio Peruzzi
is the Professor of the History of Science at the University of Padua. His main research interests are in the history of physics from the sixteenth to the twentieth centuries, foundations of quantum mechanics, and the philosophy of physics.

Tacye Phillipson
is Senior Curator of Science at National Museums Scotland, with particular expertise in the physical sciences, and twentieth and twenty-first century material.

Sara J. Schechner
is the David P. Wheatland Curator of the Collection of Historical Scientific Instruments and Lecturer on History of Science at Harvard University. She serves on the Editorial Board for the Scientific Instrument Commission's series *Scientific Instruments and Collections*. Her recent books include *Time of Our Lives: Sundials of the Adler Planetarium*, Adler Planetarium, Chicago, 2019. Her research has focused on time-finding and social change, instruments of glass, colonial astronomy, and workshop practices in America.

Sofia Talas
is the Curator of the Giovanni Poleni Museum of the History of Physics, at the University of Padua. Her research focuses on scientific instruments, their history, and interpretation.

Luìs Tirapicos
is a researcher at the Interuniversity Centre for the History of Science and Technology in the Faculty of Sciences of the University of Lisbon. His main research interests are the history of astronomy in Portugal and Spain, Jesuit science, Iberian archaeoastronomy, and the material culture of science. Currently, he is also a Contributing Editor to the *Isis Current Bibliography*, and an Executive Editor of the *HoST-Journal of History of Science and Technology*.

George N. Vlahakis
is Assistant Professor in the Hellenic Open University in Patras, Greece, and Fellow Researcher in the Institute of Historical Research/ National Hellenic Research Foundation. He is a member of several academic societies. He is also a corresponding member of the International Academy of the History of Sciences, and President of the Commission for Science and Literature of the IUHPST/ DHST. Among the subjects of his research are history of science

in southeastern Europe (eighteenth to twentieth centuries), history of scientific instruments, science and religion, science and society, and science and literature.

Jane Wess
was Senior Curator at the Science Museum, London, until 2012. During that time, she worked on the King George III collection of scientific instruments, and the mathematics collection. Following this, she undertook a PhD at the University of Edinburgh, working on the role of instruments in exploration, based at the Royal Geographical Society, London. Her interests are in scientific instruments, measurement, and the role of mathematics in science, focusing on the eighteenth and nineteenth centuries.

INTRODUCTION

Failed Historical Scientific Instruments

Sara J. Schechner and Sofia Talas

Historians and philosophers of science have long recognised that failures are an integral part of scientific practice, but science is still generally presented to the public and in schools as a one-way series of successes made of great discoveries and relevant results. Negative outcomes, discarded hypotheses, difficulties, errors, steps backward, and dead ends are barely mentioned if not totally neglected in science teaching and popularisation. In science centres and museums, exhibits also tend to showcase the instruments involved in significant scientific or technological achievements, those widely adopted, or those connected to a famous scientist or maker.

Curators of collections of historical scientific apparatus know that these displayed instruments tell only part of the story. Collection storage and archives offer examples of quirky instruments that never caught on, precision devices that never measured up, and tools that had disastrous social consequences. This book will focus on them and ask what can be learned from failures.

Of course, it is far from easy to define *failure* precisely in a scientific and technological context. Might it be a failure when a scientist tests a hypothesis by means of experiment and finds it false, or when detection instruments give inconclusive results? Perhaps. But negative and inconclusive data are valuable bricks in the construction of knowledge because they encourage researchers to try new hypotheses, build new instruments, and develop better theories. Failure is an expected part of scientific practice, and a proposition, hypothesis, or theory that lacks the capacity to be falsified or proven wrong is not considered "scientific." Even so, examples of instrumentation involved with these sorts of failures offer only a limited view of how a scientific instrument might fail.

As a category, the term *failure* must be measured with respect to something else. Just as with terms like *salty* or *tall*, *failure* demands a context. Failure with respect to what? Failure how? Failure why? If we are going to try to understand or categorise failures, we must expand our field of view and go beyond technical details. The thirteen papers in this book will do just that, showing how complex and diverse failures can be.

An obvious place to start are instruments that failed because they had flawed technical or mechanical designs, or they lacked originality and usefulness. Richard Dunn explores these in his paper on the Admiralty Compass

 | DOI:10.1163/9789004689107_002

Observatory, which was established in the 1840s in the United Kingdom in order to collect and judge all experimental and prototype devices in the field of magnetic instrumentation. Its collection, therefore, included devices that were not sufficiently precise, could not work, were useless, or unoriginal. By such criteria, the instruments would clearly seem to be failures, but Dunn stresses that the final assessment of an instrument was often the result of heated discussions and might even change over time.

No changes occurred in the assessment of the diffusion hygrometers devised in the 1940s by Heinrich Greinacher and described by Jean-François Loude in the second paper of this book. Diffusion hygrometers suffered from numerous technical drawbacks in terms of their accuracy claims not being backed up empirically, their limited temperature range, the absence of studies on how air pressure affected their operation, their lack of portability and restricted use to indoors, and the problems in manufacturing standardised porous plates. But on top of all this, the diffusion hygrometer was a commercial failure because it did not surpass prior art – the good, old, simple hair hygrometer.

Building on technical problems causing failures, Johan Kärnfelt discusses the failed equatorial mount of a Herschel reflecting telescope whose optical elements were delivered in 1790 to the Royal Swedish Academy of Sciences. Kärnfelt attributes the responsibility for this failure not only to the mount maker but also to the management of the commission by the Academy, which did nothing to rectify the situation for fifty years. Consequently, the telescope was never used for research, but only for public demonstrations and later as an exhibition showpiece.

The story of Newall's telescope presented by Panagiotis Lazos and George N. Vlahakis similarly involves a period of failure followed by success in a new arena. Built by Thomas Cooke for Robert Stirling Newall in 1869, the telescope was a technological masterpiece with superior optical and mechanical parts, but it languished unused for twenty-five years and contributed nothing to astronomy as Newall had hoped. The authors blame the failure on Newall's management of the instrument: he situated the instrument near Newcastle where the quality of seeing was compromised by pollution from factories and fog rising off the River Tyne, was too busy as an industrialist to make time for his own use of the telescope, and did not hire an observer in his place. Only after the telescope was relocated to Cambridge did it become a success, being used for cutting-edge observations by Newall's son, Hugh Frank Newall. By the late 1950s, when Newall's telescope was considered obsolete, it was relocated to the National Observatory of Athens at the expense of Greek shipping magnates. Not wanted in the UK, the 90-year-old Cooke telescope was prized by Greek astronomers doing planetary and variable-star research. After a decade or so, newer instruments replaced it, but twice more, the telescope rose from failure to new-found success – first in the 1970s in training students of physics and astronomy, and second, in the 1990s as an instrument to popularise science. Thus, the Cooke telescope's life ebbed and flowed with waves of failure and success, each measured against different standards and programmes. The papers on the Herschel and Cooke telescopes reveal the multifaceted complexity of what at first appear to be technical failures in design or geographical location.

Tacye Phillipson takes up this theme by focusing on instruments of phrenology, which are typically used as illustrations of "bad" science of the past. Should we treat the instruments of a failed science as failures themselves? Of course, phrenology had its practitioners who thought their work and tools were successful in the early nineteenth century. And today, the instruments and ideas of a failed science might be successful in exhibit settings as a means to help museum visitors better understand the methods, logic, and nuances of scientific practices that can produce dead ends as well as living branches of thought and can also sometimes recycle ideas.

The next set of papers in this volume turn from technical problems and managerial issues to the roles of social bias and behaviour – e.g., the habits and entrenched practices of the users – and commercial and economic reasons for failure.

David Bryden underscores these in his analysis of numerous mechanical logs that failed to be adopted in the eighteenth and nineteenth centuries. On one hand, they had mechanical weaknesses, were easily damaged, and were not durable enough to work at sea in rough conditions. On the other hand, they also failed because of the predisposition of the judges or purchasing agents (including those of the Board of Longitude or Royal Navy) as well as the innate conservatism of users who would rather stick with a familiar if flawed working device – as the adage says, "Better the devil you know than the devil you don't". Bryden also remarks on the commercial and economic factors that contributed to the failure of mechanical logs, such as the cost of production and pricing due to supply and demand, and the lack of investors willing to put time and money into prototyping, testing, and producing the devices on a larger scale.

Failures that arise from struggles to appeal to targeted users is the topic examined by Andreas Junk in his paper on the transmission x-ray microscope, introduced to scientists in 1967. Such a microscope showed great promise for biological research, but biologists preferred their traditional tools because the information gained by using the new microscope did not seem to justify the expense. Though a failure in the context of biology, transmission x-ray microscopes were successful in physics research. Thus, one person's failure is another person's success. Assessment depends on the time and place, and the individuals making the judgments.

Razieh Mousavi discusses a similar situation in a totally different context – the development and diffusion of *qibla*-indicators in sixteenth-century Iran within the frame of power struggles between new religious and political authorities and mathematicians and astronomers. Mathematical apparatus and methods for finding the direction of Mecca that were more precise but also more complex were deemed failures, because consumers and authorities wanted simple input-output calculators.

By analysing two cases of apparently failed schemes to read angular scales – one proposed by Pedro Nunes in 1542 and the other by Orazio Borgondio in 1720 – Luìs Tirapicos also shows how the difference between success and failure can depend on an instrument's ease of construction and use. In particular, he argues that transversal scales won out over the nonius, despite its better accuracy, because transversals were easier to make and use. Furthermore, Tirapicos underlines how the lack of marketing can contribute to make an instrument fail and be abandoned.

Michelle Mercier then sheds more light on the role of social bias in failures by examining Thomas Young's eriometer, a practical application of his optical investigations on diffraction. Though it provided precise enough results, the device failed to become established in the English medical, agricultural, or textile industries for several reasons: the poor reputation of Young as a physician, the rejection of his wave theory of light, and the steep learning curve faced by the targeted groups of users.

Focusing on the Wild photo-theodolite purchased by the Royal Geographical Society in 1925, Jane Wess adds economic causes into the social mix of reasons to fail, and she draws attention to the importance of the context in which an instrument is used. The Wild photo-theodolite was not only costly and cumbersome, but also difficult to use and repair on an expedition to the Great Barrier Reef in 1928, so that the intended user judged it a nuisance and failure. As Wess points out, an instrument that should succeed in theory can be a failure in reality as a result of difficult working conditions and the lack of logistical support to maintain it. However, after the Second World War Wild photo-theodolites became widely used, proving once again that the assessment of any instrument may radically change with time.

Besides social and economic factors, politics can also contribute to make an instrument become a failure, as Sofia Talas and Giulio Peruzzi show in their paper on the cloud chamber made in Padua in the 1930s by Bruno Rossi, one of the pioneers of cosmic-ray research. Due to fascist racial laws (which forced Rossi, who was Jewish, to leave Italy) and the advent of the Second World War, the cloud chamber was not used until the end of the war, by which time it was obsolete and could produce no significant scientific results. It could thus be seen as a failure for cutting-edge research. On the other hand, the authors

FIGURE 0.1
This young lady finds her binoculars very unsatisfactory and might blame their optical quality, but in context, the causes of their failure include user errors and environmental factors. Victorian trade card, likely American
SCHECHNER COLLECTION

point out that it had great success as a "training for research" device at a time when Italian physics was trying to emerge from its post-war difficulties.

Thus far, a reader might be convinced that there are numerous ways in which a historical scientific instrument can fail and that each should be understood in a nuanced context (*Fig. 0.1*). The closing paper of the volume challenges this point by offering us the perspective of an instrument maker. Nicolàs de Hilster argues that failure lies not in the apparatus but in the process of its production. What failed in a "failed instrument" were the designer and fabricator, who did not go far enough in creating a tool that solved the problem that it set out to do. Failure is not about things in context, but the method and practice of making them.

So where does this leave us? The papers collected in this book illustrate how tricky it is to fully understand what separates success from failure. Success and failure are not poles of an on-off switch. This volume, we hope, breaks new ground by its focus on failed historical scientific instruments, because their stories offer many insights into the development of science and technology and the social factors involved. Failed instruments are worthy of preservation, study, and exhibition.

CHAPTER 1

An Archive of Failed Inventions: the Museum of the Admiralty Compass Observatory

Richard Dunn

Failure is a matter of judgment. There are also many ways in which an invention might be judged to have failed, whether in theory, practice, or originality. This chapter explores the ways in which a specialist museum created by a branch of the British Royal Navy collected and examined magnetic compasses in order to assess inventions, judging many as failures in the process. It was a function that began with the committee that recommended the creation of the Admiralty Compass Department, and continued in the later work of what became the Admiralty Compass Observatory (ACO). As well as being at the forefront of compass development, the ACO became a site at which the success and failure of one type of navigational instrument was judged, with the uses of its museum holdings demonstrating the value of failed instruments. As the example of Henry Constantine Jennings's insulating compass will show, moreover, failure was not permanently fixed but subject to changes in theoretical understanding.

1 The Admiralty Compass Observatory and Its Museum

Although the magnetic compass had been used at sea for centuries, it was well known by the 1830s that it was likely to err as a navigational instrument. Following a suggestion in 1837 by Francis Beaufort, Hydrographer to the Navy, the British Admiralty formed a committee to look into and improve what the invitation described as "an evil so pregnant with mischief".[1] The resulting body, the Admiralty Compass Committee, comprised six men known for their knowledge and experience in magnetic matters:

- Captain James Clark Ross, a renowned naval officer and explorer, chairman of the Committee until 1839, when he took command of an expedition to the Arctic

1 Letter to prospective members of the Admiralty Compass Committee, 18 July 1837, quoted in A. E. Fanning, *Steady as She Goes. A History of the Compass Department of the Admiralty*, Her Majesty's Stationery Office, London, 1986, p. 3.

 | DOI:10.1163/9789004689107_003

- Hydrographer Francis Beaufort, who took over as chairman from 1839
- Major Edward Sabine, a geophysical expert with experience on Arctic expeditions
- Samuel Hunter Christie, a leading investigator into magnetic phenomena, who also taught at the Royal Military Academy in Woolwich from 1838
- Commander Edward John Johnson, a Royal Navy officer who had conducted magnetic experiments for the Admiralty in 1835
- Captain Thomas Best Jarvis, an experienced surveyor, formerly of the Bombay Engineers, who remained on the Committee until he left for India in 1838

Between 1837 and 1840, the Committee investigated compass construction by examining examples from Britain and overseas and consulting with other compass experts, ultimately proposing a new design, the Admiralty Standard Compass.[2] An important additional outcome was the formation of the Admiralty Compass Department in 1842, with Edward Johnson appointed as Superintendent of Compasses. Two years later, the Department acquired a home in Maryon Road, Charlton, near Woolwich Dockyard, with an observatory for compass testing and research in the large garden, hence its official address, Admiralty Compass Observatory, Woolwich.

Importantly for this chapter, the need for a dedicated museum for the compasses assembled by the Admiralty Compass Committee was recognised from the start. As a later catalogue states,

> such a collection would be valuable from both the historical and technical points of view, and it was added to from time to time as opportunity offered, with the result that by the end of the 19th Century it had assumed considerable dimensions.[3]

The ACO moved to Deptford Dockyard in 1869 and then to Ditton Park, near Slough, in 1917. Both sites had provision for the ever-expanding collection of compasses and related material, most of which passed to the National Maritime Museum, Greenwich, shortly before the Ditton Park site was sold in the mid-1990s.

The first catalogue of the ACO collection was created in about 1850 by James Brunton, a pensioner sergeant of Artillery who had been appointed as

2 Fanning, *op. cit.* (n. 1), pp. 3–6; for examples, see Admiralty Standard Compass card, 1840, National Maritime Museum, Greenwich (NMM), museum no. ACO0962, and Admiralty Standard Compass, 1842, NMM, museum no. ACO1445.

3 Lieutenant G. H. Alexander (with additions by Commander W. E. May), "Admiralty Compass Observatory Museum, Catalogue and Notes", 1907–1951: London, NMM, ACO/S&T/17/8.

Compass Examiner in September 1844.[4] The catalogue's arrangement and contents (*Table 1.1*) reveals the intent of the collection, notably in recording of date of invention or of being received (i.e., a known date by which the idea had been realised), and in the notes on construction and use. Each of these columns held information that could aid the assessment of originality of invention and functional capability.

TABLE 1.1 The first fifty entries of "A List of Compasses, Instruments &c. at the Compass Observatory", about 1850

No.	Description of instrument		Name of maker or inventor	Date of invention or of being received	From whom received		Remarks on the construction of the instrument and its intended use	
1	Chinese compass		- -	1848	Mr Brunton			
2	Hanging	*do*	Bradford & Son, Lon[do]n	"	Hydro[graphic] Office		Metallic cap and pivot	
3	*do*	*do*	Marked II. 104	"	*do*	*do*	*do*	*do*
4	*do*	*do*	Stebbing Portsmo[uth]	"	*do*	*do*	Agate cap steel pivot	
5	*do*	*do*	Syeds Rother[hi] the	"	D[ock] Y[ard] Wool[wic]h		*do*	*do*
6	Boats *do*		No name	"	*do do*		Ex[am]ple suspended on gimbals in brass bowl	
7	*do*	*do*	*do*	"	*do*	*do*	*do* Bowl upon a pivot suspended in gim[ba]ls	
8	*do*	*do*	*do*	"	Hydro[graphic] Office		Fitted with sight vane	
9	*do*	*do*	Gilbert, London	"	*do*		Experimental	
10	*do*	*do*	*do*	"	*do*		*do*	
11	*do*	*do*	*do*	"	*do*		*do*	
12	*do*	*do*	No name	"	*do*		Card suspended in double gimbals	

4 James Brunton, "A List of Compasses, Instruments &c. at the Compass Observatory", about 1850: London, NMM, ACO/13/24.

TABLE 1.1 The first fifty entries of "A List of Compasses, Instruments &c. (*cont.*)

No.	Description of instrument	Name of maker or inventor	Date of invention or of being received	From whom received	Remarks on the construction of the instrument and its intended use
13	*do* *do*	Preston, London	”	D[oc]k Y[ar]d Woolwich	*do* and fitted with heavy needles
14	*do* *do*	*do* *do*	”	*do* *do*	*do* *do* in brass binnacle
15	*do* *do*	*do* *do*	”	*do* *do*	Liquid, fitted with feeding reservoirs, needle cased
16	*do* *do*	Barrow	”	Hydro[graphic] Office	Exp[erimenta]l card in double gimbals, 4 compound needles
17	*do* *do*	*do*	”	*do* *do*	*do* ruby pivot, metal cap. stops on card, centre pin
18	Binnacle *do*	Blunt, New York	1840	D[oc]k Y[ar]d Woolwich	Wood bowl, paper card, metal cap (very old)
19	*do* *do*	McAuly, Dublin	”	*do* *do*	*do* *do*
20	*do* *do*	Sir W. S. Harris	”	*do* *do*	Pivot stem fitted with double gimbals
21	*do* *do*	Pope	”	*do* *do*	Single dipping needle
22	*do* *do*	Mr George, R.N.	1849	Mr George	Needle curved in the direction of the dip
23	*do* *do*	Captn. Kater	1848	Hydro[graphic] Office	Double suspension, silk fibre & pivot
24	*do* *do*	Preston	1844	*do* *do*	Card moving in a vertical axis on two centres
25	*do* *do*	*do*	”	Mr Dent	*do* *do* *do*
26	*do* *do*	Dent's Axis	”	D[oc]k Y[ar]d Woolwich	Compass set on four glass pillars

TABLE 1.1 The first fifty entries of "A List of Compasses, Instruments &c. (*cont.*)

No.	Description of instrument	Name of maker or inventor	Date of invention or of being received	From whom received	Remarks on the construction of the instrument and its intended use
27	*do* *do*	Crowe, Faversham	1820	*do* *do*	Liq[ui]d az[imuth] card enameled on copper, boyant, semisperical [sic.]
28	*do* *do*	Thomson, London	"	*do* *do*	Point of suspension one inch above plane of card
29	*do* *do*	Miller, Devonp[or]t	1848	Devonport	Needle construction in the line of the dip
30	*do* *do*	French	"	D[oc]k Y[ar]d Woolwich	Pivot in card, working on agate plane, moveable
31	*do* *do*	Stebbing, Portsm[outh]	"	*do* *do*	Transparent, silk flies, lighted from main deck
32	*do* *do*	Sir E Owen	1840	*do* *do*	Bowl suspended on spiral Springs
33	*do* *do*	*do* *do*	"	*do* *do*	Double gimbal card by Dickman of Leith
34	*do* *do*	*do* *do*	"	*do* *do*	*do* Thos. Jones, London
35	*do* *do*	*do* *do*	"	*do* *do*	Common suspension, Syeds, Rotherhithe
36	*do* *do*	Sir W. S. Harris	1848	*do* *do*	Double inverting pivot, and inverting cap
37	*do* *do*	Mr St. John. U.S.A.	1851	Devonport	Two small needles above the plane of the card
38	*do* *do*	Lt. Edye R.N.	1848	*do*	Needle suspended to pole of a vertical magnet

TABLE 1.1 The first fifty entries of "A List of Compasses, Instruments &c. (*cont.*)

No.	Description of instrument		Name of maker or inventor	Date of invention or of being received	From whom received	Remarks on the construction of the instrument and its intended use
39	*do*	*do*	Mr Barrow	"	D[oc]k Y[ar]d Woolwich	Elliptic needle, bowl suspended on a pivot
40	*do*	*do*	Captn. Phillips RN	1840	Mr Barrow	Bowl in double gimbals in padded box
41	*do*	*do*	Mr Keen, Liverpool	1854	Mr Keen	Porcelain bowl, pivot resting on springs
42	*do*	*do*	Mr Gowland *do*	"	Gowland	Vertical card protected by glass bowl
43	*do*	*do*	Jennings, London	1835	D[oc]k Y[ar]d Woolwich	Needle surrounded by Iron or steel filings
44	*do*	*do*	Wood, Liverpool	1830	*do do*	Pointed needle fitted to a revolving vertical circle
45	*do*	*do*	Lt. Kortright R.N.	1854	Lt. Kortright	Cast iron bowl
46	Boats compass		Comdr Walker RN	1848	D[oc]k Y[ar]d Wool[wic]h	Suspension in imitation of Thompsons (No. 28)
47	Azimuth	*do*	No name	"	Hydrographic Office	Old land compass
48	*do*	*do*	Blunt London	"	Hydro[graphic] Office	Common az[imuth] old pattern
49	*do*	*do*	Schmalcalder *do*	"	*do* *do*	Prism & adjustments on Schmalcalder's principle
50	*do*	*do*	Gilbert *do*	"	*do* *do*	Experimental

NATIONAL MARITIME MUSEUM, GREENWICH, ACO/13/24

2 Testing Success and Originality

The idea of assembling a reference collection to assess technical details and claims to originality began with the work of the Admiralty Compass Committee, which regularly referred to items in its possession. In discussions in 1839 with the Arctic explorer and magnetic expert, William Scoresby, for example, the Committee drew on examples including a Danish compass to challenge the originality of Scoresby's method for producing strongly magnetised needles.[5] Scoresby strongly objected to their assessment, however, insisting that,

> The principle of construction … by different wires which was shewn me by the Committee in (I think) a Danish Compass, does not at all invalidate the newness and originality of my needles – for in that formation of bars or needles in laminae & these tempered or hardened throughout their extent, consists the value of the discovery.[6]

Committee member Edward Sabine responded that Scoresby "had to learn that the track which he had been pursuing was by no means the untrodden one that he imagined", since they also had another compass that used Scoresby's method:

> In the form which Mr. Scoresby specifies, namely '*combined plates*', it so happens that the Committee have in their possession a compass made by Gilbert, certainly prior to 1823, which perfectly answers this description.[7]

Fellow Committee member Samuel Hunter Christie agreed that either the Gilbert or the Danish compass invalidated Scoresby's claim:

> If then Mr. Scoresby claim as a discovery, the employment of a needle composed of laminas of steel, here [in Gilbert's compass] is this needle which undoubtedly existed 13 years before his was exhibited. If he claim, the keeping of the needles quite separate, as a discovery, there are the

5 Presumably *ibid.*, no. 56; Alexander, *op. cit.* (n. 3), no. 514; NMM, museum no. ACO0467.

6 William Scoresby, letter to James Clark Ross, 1 February 1839, in "Report of the Admiralty Compass Committee": London, NMM, ACO/11/6, p. 123.

7 Edward Sabine, letter to Ross, 26 February 1839, in "Report of the Admiralty Compass Committee", 1840: London, NMM, ACO/11/6, p. 139.

> Danish compass, Gilbert's compound needle, and also the experiments made with it and other compound needles.[8]

This sort of argument would be particularly important because the Admiralty Standard Compass recommended by the Committee included a compound needle similar to that in the Danish example. The Danish compass was material evidence, therefore, that the idea was already well known and not a recent invention subject to proprietary interests.

With the ACO established, its museum continued to function as an archive for the comparison and testing of new ideas. This included keeping an eye on developments overseas. Visiting in 1850, journalist Frederick Hunt reported after visiting the ACO that,

> With the open liberality that unites in brotherhood the scientific men of all countries, France and Denmark sent specimens of what their best men had succeeded in perfecting for the use of their navies. These instruments are very good, and attract deserved attention in the observatory-collection of specimens. […] The Dane is a good substantial instrument, even more excellently finished than the compasses issued in our navy.[9]

Failures had a place too and were documented as such. Pope's dipping needle compass of 1820, for example, was found to be "practically useless when the needle lay in the line of dip".[10] Nevertheless, it was evidence of an idea having been tried at a certain date as well as of the failure of that idea in practice. As an archive of inventions, the ACO Museum could thus be used to test whether a specific design was truly original. Indeed, Hunt revelled in the imagined disappointment of the hopeful inventor:

> A sanguine theorist completes what to him is perfectly new. Certain that he is to be immortalised and enriched, he sets off to the Observatory with his treasure, to reveal his grand secret, and receive the anticipated reward. He is shown into the compass-room, and there, – horror of

8 Samuel Hunter Christie, letter to Ross, 8 March 1839, in "Report of the Admiralty Compass Committee", 1840: London, NMM ACO/11/6, p. 141; see also Jennifer Bulstrode, "The Eye of the Needle: Magnetic Survey and the Compass of Capital in the Age of Revolution and Reform", unpublished doctoral thesis, University of Cambridge, 2019, pp. 168–173.

9 Frederick Knight Hunt, "'Swinging the Ship': A Visit to the Compass Observatory", *Household Words* 1 no. 18 (27 July 1850), pp. 414–418: 417.

10 NMM, museum no. ACO0008; Alexander, *op. cit.* (n. 3), cat. no. 11; see also Table 1, no. 8.

> horrors, – upon the table, amidst a host of others, there is an old discarded instrument the very counterpart of his own! It was made, and tried, and discarded, years ago.[11]

It may be that Hunt heard of Scoresby's dealings with the Admiralty Compass Committee during his visit and imagined this scenario. Whether this is the case or not, it is clear that by the mid-century the ACO saw the value of its museum in judging the claims to originality of compass inventors.

3 Fail, Fail Again

The case of a compass invented by Henry Constantine Jennings, a manufacturing and consulting chemist as well as an inventor, shows both how the assessment of an instrument might change over time and the ACO Museum's continued function as a reference archive in the decades after its creation.

Jennings contacted the Admiralty in 1815 with "a method of making a Compass in such a manner as always to point *Due* North, without any Variation". He claimed to have invented an "Insulating Compass" that was immune to the deranging effects of a ship's iron and other local sources of magnetism yet still able to respond to the Earth's magnetic field. This, he claimed, would mean that the mariner could determine latitude and longitude with the compass and appropriate tables.[12] A couple of years later, Jennings also approached the Board of Longitude, which could still reward improvements in navigation.[13]

Jennings patented his design in November 1818 (*Fig. 1.1*). The patent (no. 4259) describes the innovation as "a means of guarding or protecting the magnetic needle ... from all action arising from iron in its neighbourhood by the addition of curved pieces of iron to the underside of the card" (*Fig. 1.2*).[14] These curved pieces,

---

11 Hunt, *op. cit.* (n. 9), p. 415.

12 Henry Constantine Jennings, letter to the Board of Admiralty, 15 May 1815: London, The National Archives (TNA), ADM 1/4782, no. 141.

13 Henry Constantine Jennings, letter to the Secretary of the Board of Longitude, August 1817: Cambridge, Cambridge University Library (CUL), RGO 14/31, fols. 204–205; Richard Dunn, "The Impudent Mr Jennings and His Insulating Compass", *Bulletin of the Scientific Instrument Society*, no. 111 (2011), pp. 6–9.

14 The patent shows four guards (*Fig.* 1.2), although a surviving example in the NMM (museum no. ACO1517) has only two (*Fig.* 1.4).

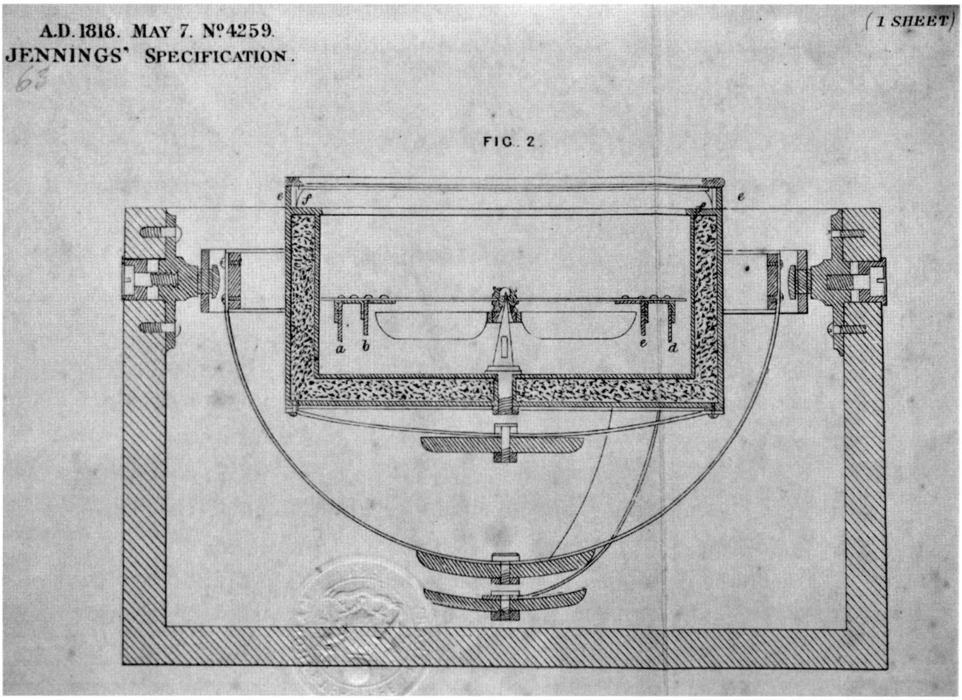

FIGURE 1.1 Patent drawing of the Jennings compass, showing (roughly stippled) the space for the treated iron filings. From Henry Constantine Jennings, *Specification of Henry Constantine Jennings. Mariners' Compass.* Eyre and Spottiswoode, London, 1857

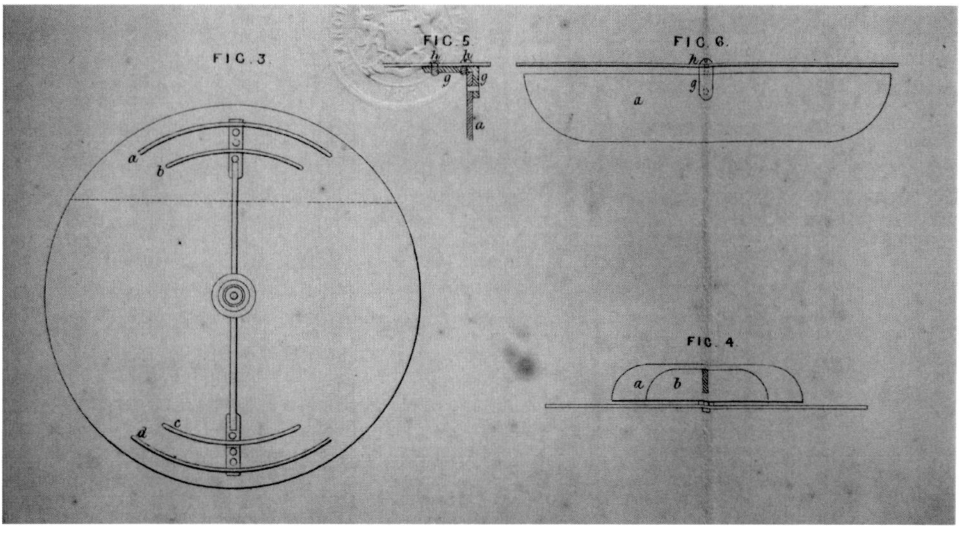

FIGURE 1.2 Patent drawing of the card of the Jennings compass, showing the iron "shields" (a, b, c, and d). From Henry Constantine Jennings, *Specification of Henry Constantine Jennings. Mariners' Compass.* Eyre and Spottiswoode, London, 1857

> act as guards against the passage of the magnetic fluid, by absorbing the first quantity of it, the remainder being absorbed by means of iron filings … in the space between the sides of the brass or copper boxes which enclose the needle and card.[15]

The efficacy of the magnetic shielding relies on the correct preparation of the iron pieces and iron filings to make them as pure as possible and free of carbon (which, Jennings says, will render the iron prone to becoming polarised). If the correct procedure is followed,

> two inches and a half thickness will, I conceive, insulate from one ton of iron close to the compass box, and from one hundred tons if at one foot distance, and so on in proportion as the distance increases.[16]

In a pamphlet sent to the Board of Longitude, Jennings claimed that his method represented nothing less than a "new discovered Law of Nature".[17]

Correspondence with the Admiralty, the Navy Board and the Board of Longitude lasted initially until the 1820s and included several sea trials. While tests on HMS *Tonnant* in 1818 concluded that Jennings's design did not have "any particular advantage" over compasses by Alexander, Stebbing and Dunsterville, Rear Admiral Hallowell noted that it would be "a great point gained" if its insulating property could be established.[18] Lieutenant Reneau of the *Tonnant* was more enthusiastic, believing that the principle underlying Jennings's method was sound, having had the inventor demonstrate it to him.[19] Trialling Jennings's compass on HMS *Isabella* in search of a North-West Passage in 1818, Captain John Ross found that it "completely obviated the effect of local attraction", although it "ceased to act when the variation was great".[20] George

---

15 Henry Constantine Jennings, *Specification of Henry Constantine Jennings. Mariners' Compass* [Patent No. 4259, 7 November 1818], Eyre and Spottiswoode, London, 1857, p. 2.

16 *Ibid.*, p. 3.

17 Henry Constantine Jennings, "Mr. Jennings's Observations on the Improvement of the Mariner's Compass, by the Insulation of the Needle", *c.*1817: Cambridge, CUL: RGO 14/31, fol. 203.

18 Rear Admiral Hallowell, letter to the Admiralty, 2 August 1818: London, TNA: ADM 1/630.

19 Lieutenant Henry Reneau, "Report on Jennings' Insulated Compass", 1 August 1818: London, TNA: ADM 1/630.

20 John Ross, *A Voyage of Discovery Made Under the Orders of the Admiralty, in His Majesty's Ships Isabella and Alexander, for the Purpose of Exploring Baffin's Bay, and Inquiring into the Probability of a North-West Passage*, John Murray, London, 1819, Appendix, p. cxxv; see also Richard Dunn, "North by Northwest? Experimental Instruments and Instruments of Experiment", in Fraser MacDonald and Charles W. J. Withers (eds.), *Geography, Technology and Instruments of Exploration*, Ashgate, Farnham, 2015, pp. 57–75: 64–66, 70–72, 75.

Fisher, astronomer on HMS *Dorothea*, which sailed with HMS *Trent* towards the North Pole as part of the same Arctic expedition, formed a similar judgment, writing that the Jennings compass "appears to resist the action of iron *not magnetical*", although it was equally prone to the errors observed in other compasses.[21]

Jennings, however, claimed the good opinion of over 2500 sailors and could point to Admiral Penrose's judgement that his insulating compass was "the most important invention that he had ever seen".[22] Nevertheless, the inventor's claims and reports of apparent success were not sufficient to convince either the Admiralty or the Board of Longitude. Furthermore, his aggressive communications did little to ingratiate such an "impudent Fellow" to them.[23]

The ACO, which had acquired its Jennings compass in 1835 (see *Table 1*, no. 43), became embroiled in the matter in 1846, when Jennings approached the Admiralty once more after reading a letter on compass deviation in the London newspaper, *The Times*.[24] Captain Johnson, still heading up the ACO, was now quite clear on the matter:

> There have been numerous proposals for intercepting the *local* influence of Iron upon the Compass, but on Submitting them to the proper test they as might have been expected signally failed. […]
>
> So far as our present knowledge extends there is no substance in nature which will cut off the action of one magnet upon another in given positions.[25]

In January 1847, Johnson added that he had discovered a compass marked "Jennings Insulating Compass" (*Figs. 1.3* and *1.4*) at the ACO, put it "to the most obvious test, and find that the magnetic needle is not shielded from the external influence of Iron". The compass was then at the Admiralty, he added, so "a repetition of the fact may be seen by direct experiment at any time".[26]

21 George Fisher, "On the Variation of the Compass, Observed in the Late Voyage of Discovery to the North Pole, under Captain Buchan, and on the Errors in Longitude, Produced by the Action of Iron in Ships upon Chronometers", *Quarterly Journal of Science, Literature and the Arts*, IX (1820), pp. 81–106: 81.

22 James Horsburgh, "On the Insulated Safety Compass Lately Invented by Mr Jennings", *Philosophical Magazine*, 53 (1819), pp. 365–367: 366.

23 Sir John Osborn, letter to Sir Joseph [possibly Sir Joseph Sydney Yorke], 30 November 1817: Cambridge, CUL RGO 14/31, fol. 210; Dunn, *op. cit.* (n. 13), p. 8.

24 Presumably "An Officer in a Steamer", "Local attraction", *The Times*, 5 December 1846, p. 7.

25 Captain Johnson, letter to Admiral Beaufort, 11 December 1846: London, NMM, ACO/13/1.

26 Captain Johnson, letter to Captain Hamilton, 22 January 1847, in "Transcripts of Manuscripts from the National Archives": London, NMM, MAA/19. The ACO's Jennings compass is NMM, museum no. ACO0517.

FIGURE 1.3 Magnetic compass inscribed "insulating Compass No 6" and "Jennings & Cos. Patent", London, about 1818
NATIONAL MARITIME MUSEUM, GREENWICH, AC01517. PHOTO © NATIONAL MARITIME MUSEUM, GREENWICH

FIGURE 1.4 Underside of the card of the Jennings compass with two iron "shields", London, about 1818
NATIONAL MARITIME MUSEUM, GREENWICH AC01517. PHOTO © NATIONAL MARITIME MUSEUM, GREENWICH

By 1847, then, the Jennings insulating compass could be unequivocally judged a failure in theory and in practice. In terms of the theoretical assessment, magnetism, variation, and deviation had been poorly understood in the early nineteenth century when Jennings first offered his invention. In particular, there was no consensus as to the causes of magnetic deviation. Writing about the deviation of compasses on the Arctic voyage he commanded, for example, John Ross postulated several factors, including temperature, wind, the ship's heading, magnetic variation and dip.[27] What counted most in the trials around 1818, therefore, was whether the Jennings compass appeared to work as claimed. Some thought it did; others were more circumspect. By the time of Johnson's damning assessment in the 1840s, theoretical and practical work, including that of Siméon Denis Poisson to develop mathematical theories of magnetic deviation and of Matthew Flinders and Peter Barlow on practical solutions to compensate for it, had changed practitioners' understanding.[28] Johnson could now be confident in dismissing Jennings's ideas in principle, reinforcing his theoretical understanding by using an instrument from the ACO Museum to demonstrate the failure of the insulating compass in practice.

This was not the end of the matter. In 1867, a letter arrived from the Board of Trade for Captain Evans, Surveyor of Compasses since 1855. It asked if he knew anything of Jennings, who had become "very savage" due to the Board's reluctance to reopen the question of adopting his compass. Evans's reply could not have been clearer:

> In the opinion of the authorities at that time [the 1840s] the compass was an utter failure and the Hydrographer (the late Sir F. B. [Francis Beaufort]) had to complain of the dogmatic and presumptuous tones of Dr Jennings letters […]
>
> It appears that old age has not softened the Dr Jennings style of letter writing or developed his knowledge of magnetic science. I believe his Insulating Compass to be an entire delusion.[29]

27 Ross, *op. cit.* (n. 20), Appendix, pp. xxx–xxxi.

28 Charles H. Cotter, "George Biddell Airy and his Mechanical Correction of the Magnetic Compass", *Annals of Science* 33 (1976), pp. 263–274: 263–265; Charles H. Cotter, "The Early History of Ship Magnetism: The Airy-Scoresby Controversy", *Annals of Science* 34 (1977), pp. 589–599: 589–591.

29 Captain Evans, letter to Board of Trade, 18 September 1867, in "Transcripts of manuscripts from the National Archives": London, NMM, MAA/19.

4 Conclusions

Material examples of instrumental failure in the Museum of the Admiralty Compass Observatory had specific uses: to investigate their operation and efficacy; to make comparisons with other proposals; and to assess claims of originality in invention. These functions were put in place as the museum collection was formed in the nineteenth century and continued well into the twentieth century. Thus in 1939, when the British Army trialled a tank compass designed by Peter Schilovsky, its inventor was apparently furious when told that,

> the idea had been proposed by Captain Milne in 1832 and by Captain Sparks of the P&O about 1848 and found useless [...] Nevertheless, he took out a patent [...] and claimed a reward because compasses had been fitted in tanks subsequently. The claim was heard in 1947 and disallowed.[30]

Failed ideas and instruments were research tools and sources of evidence and could remain so for many decades. For historians of instruments, such failures and the stories around them serve equally as valuable evidence of the process of invention and adoption (or not), of the means by which invention has been assessed and of some of history's forgotten figures. All these, after all, are part of the story of the development of instrumental technologies.

30 NMM, museum no. ACO0681; Alexander *op. cit.* (n. 3), cat. no. 1901.

CHAPTER 2

The Diffusion Hygrometer: a Commercial Failure

Jean-François Loude

Louis Dufour, professor at the Academy of Lausanne (Switzerland), observed in 1874 a pressure difference across a porous wall separating moist air and a vessel containing either water or a desiccant, pointing to a possible design for a hygrometer.[1] A long period of almost complete oblivion followed until 1944 when Heinrich Greinacher (1880–1974), then physics professor at the University of Bern (Switzerland) and Director of the Meteorological Observatory of that city, took a sudden interest in instruments to measure the atmosphere's moisture content. The performances of hygrometers then in common use were certainly not satisfactory. Greinacher rediscovered the work of Dufour, claimed to build the first diffusion hygrometer, proposed a theory, and even got a patent for it. In the course of the 1950s, two different models were produced by the young firms of Lausanne, Rüeger, and TESA.

We found one exemplar of each maker in the Physics Museum of the University of Lausanne (UNIL) and the École Polytechnique Fédérale de Lausanne (EPFL) and tested them.[2]

These diffusion hygrometers were ultimately a commercial failure. Their drawbacks were too numerous, and they soon became superseded by novel electrical and electronic types of instruments.

---

1 This chapter is based on my identically named, more detailed presentation at the XXXIX Scientific Instrument Symposium, London, UK, 14–18 September 2020. See: https://www.academia.edu/44161411/The_Diffusion_Hygrometer_A_commercial_failure (accessed 17 August 2023).

2 The Collection of Scientific Instruments UNIL-EPFL – dubbed the Physics Museum – at Lausanne, contains more than a thousand objects spanning nearly three centuries of teaching and research. Since 2003, I have been making an inventory of the collection. The full inventory is now available online (in French). The texts and images are placed under a Creative Commons license and can be reused freely. For a virtual visit, go to https://www.epfl.ch/campus/art-culture/museum-exhibitions/collection-of-scientific-instruments/ (accessed 17 August 2023).

 | DOI:10.1163/9789004689107_004

1 Diffusion through Porous Materials

Louis Dufour was born in 1832 near Montreux (Canton de Vaud, Switzerland). After studying physics at Lausanne, then at Paris, he became in 1855 professor at the Academy (later University) of Lausanne.[3] For unknown reasons, he declined invitations to the Polytechnikum of Zurich (1867) and to the University of Bern (1868). He was forced to retire because of physical illness in 1878 and died in 1892. His successor was Henri Dufour (no kin).

For his research, Louis Dufour had a very modest laboratory, in a small unheated room without running water behind an auditorium in the Academy building. He had practically no money for research. The only help he got was from a student-assistant. He was interested mainly in heat-related problems (delays in the freezing of water when cooled, in its boiling when heated; the "Foehn" problem; etc.).[4] In his last experimental work, he studied the diffusion between dry air and moist air through a thin wall of baked clay, observing a difference of pressure (Δp) between the moist atmosphere of the laboratory and a vessel containing either a desiccant ($\Delta p < 0$) or water ($\Delta p > 0$).[5] He called this phenomenon *hygrometric diffusion*. In a long series of experiments, he tested different materials (such as baked clay, marble, and diverse forms of porous carbon), varying the thicknesses and sizes.

By age 46, Dufour was seriously ill and had no time to apply his results to the construction of a hygrometer. Very few scientists mentioned his work following his retirement. Among those who did, August Kundt compared existing theories of diffusion with the experimental results of Dufour, Merget, Violle and others.[6] F. Schidlowsky, in Russia, repeated the experiments of Dufour. He is reported to have built the first diffusion hygrometer, which was shown in a paper of 1886.[7] And a very short description of a diffusion hygrometer was

3 Henri Dufour, "Louis Dufour (1832–1892) – Notice biographique", *Bulletin de la Société Vaudoise des Sciences Naturelles* 76 (1893), pp. 211–228.

4 The "Foehn" is a dry, warm, down-slope wind that occurs in the lee (downwind side) of a mountain range, here in Switzerland with moist winds of the Mediterranean Sea blowing over the Alps, often tempestuously.

5 Louis Dufour, "Sur la diffusion entre l'air humide et l'air sec, à travers une paroi de terre poreuse", *Comptes rendus hebdomadaires des séances de l'Académie des Sciences, Paris* 78, no. 1 (1874), pp. 961–964; L. Dufour, "Recherches sur la diffusion entre l'air sec et l'air humide à travers une cloison de terre poreuse", *Bulletin de la Société Vaudoise des Sciences Naturelles* 13, no. 72 (1874–1875), pp. 165–217; L. Dufour, "Sur la diffusion hygrométrique", *ibid.* 13, no. 74 (1874–1875), pp. 608–641.

6 A. Kundt, "Zur Erklärung der Versuche Dufour's und Merget's über die Diffusion der Dämpfe", *Annalen der Physik*, 3rd series, 2 (1877), pp. 17–24.

7 F. Schidlowsky, "Versuch einer Anwendung der Diffusion der Gase und Dämpfe durch poröse Körper zur Bestimmung der Feuchtigkeits-und Kohlensäuregehaltes im umgebenden

given by A. Jouveneau, in Belgium, along with a drawing, showing an instrument looking more like a demonstration model than a practical hygrometer.[8] I have not been able to find anything more published between 1895 and 1944 – nearly half a century.

2 State of Hygrometry around 1950

Air moisture content is one of the three fundamental parameters describing the state of atmosphere. They are the *atmospheric pressure* (p_{atm}), the *temperature* (T), and either the *water vapour partial pressure* (p), or the *relative humidity* (RH). The relative humidity RH, measured in percent, is the ratio between the partial water pressure p of the water and the maximum possible pressure, that is to say the *saturation pressure* (p_s), at temperature T.[9] In moist air, a certain amount of water vapour replaces air, in such a way that the atmospheric pressure remains constant ($p_{atm} = p + p_{air}$). Consequently, moist air is lighter than dry air. The *absolute humidity* can be defined as the mass of water contained in a kilogram of moist air.

The measurement of the air-moisture content, the subject of hygrometry, is fundamental for meteorology and very important for human comfort, for the preservation of materials in museums, libraries, film archives, and in many industrial and commercial domains. In 1950, the instruments then available for measuring relative or absolute humidity, the hygrometers, had all been invented at the end of the eighteenth or at the beginning of the nineteenth centuries.[10] For the routine measurement of the moisture content of air, either outdoors or indoors, two types of instruments were in common use: the hair hygrometer (dating from 1783), and several models of wet-and-dry-bulb

---

Medium", *Beiblätter zu den Annalen der Physik und Chemie* 11 (1887), pp. 618–620; full paper in Russian in: *Journal der russischen physikalisch-chemischen Gesellschaft* 18, no. 6 (1886), pp. 182–204; Schidlowsky, "Attempt to apply the diffusion of gases and vapours through porous bodies to determining the amount of moisture and carbonic acid in the surrounding", *London, Edinburgh and Dublin Philosophical Magazine* 25, no. 152 (1888), pp. 77–78.

8 A. Jouveneau, "Nouvel hygromètre à diffusion", *Bulletin de la Société Belge d'Astronomie* (1895–1896), pp. 27–28.

9 Water vapour is far from being a perfect gas, but its saturation pressure p_s (measured in kilopascals) depends only on the temperature T (in °C), increasing rapidly, nearly exponentially, with the temperature.

The experimental data can be approximated by several formulas, one of the best known being the Magnus / Tetens formula: $p_s = 0.61078 \cdot \exp\,[17.27\,T\,/\,(T + 237.3)]$.

10 W. E. Knowles Middleton, *Invention of the Meteorological Instruments*, Johns Hopkins Press, Baltimore, 1969.

hygrometers (dating from about 1825). For precision measurements in the laboratory, two methods based on sound physical principles were available: either the use of several types of condensation (dew point) hygrometers (dating from 1829), or the gravimetric method (by which water contained in a given volume of air was absorbed and the absorbing material weighed). The gravimetric method remained for a long time the most accurate method.

The state of the art in 1950 was reviewed by Wexler and Brombacher.[11] Typical practical instruments were described in contemporary books, such as the one by Reiner.[12] No instrument was fully satisfactory. Novel or greatly improved types of hygrometers were clearly needed, but progress was slow, due to technological limitations. By contrast, satisfactory, easy-to-use instruments were available to measure atmospheric pressure (barometers) and temperature (thermometers), accurate to better than 0.1%.

2.1 *Hair Hygrometers*

Before 1783, hygroscopes demonstrating a dimensional change of water absorbing substances had already been built, but they had not been calibrated. In 1783, Horace-Bénédict de Saussure (1740–1799), citizen of Geneva, published his *Essais sur l'hygrométrie*. He argued, contrary to the opinion of his compatriot Jean-André Deluc (1727–1817), that human hair was the best hygroscopic material for hygrometers. Jacques Paul (1733–1796) built for him beautiful instruments, which survive in collections such as the Musée d'histoire des sciences of Geneva.[13]

Saussure made hygrometry a quantitative science, by doing extensive experiments on the vapour pressure of water at various temperatures, and publishing tables of humidity in various gases at different temperatures.

The hair hygrometer was simple and operated silently. The readout of *RH* (%) was immediate, although the scale was non-linear. It was a passive device that did not need a power supply and was easily transported. The maintenance was minimal (aside from periodic calibration). Recording models (hygrographs) were available. Their main drawbacks were the needed calibration and the low accuracy (with errors being within a few percent). Today, cheap models, using synthetic fibres instead of hairs, are still made, sold and in current use.[14]

---

11 Arnold Wexler and W. G. Brombacher, *Methods of Measuring Humidity and Testing Hygrometers*, National Bureau of Standards Circular 512 (28 September 1951), U.S. Department of Commerce, National Bureau of Standards, Washington, DC, 1951.

12 Josef Reiner, *Die meteorologischen Instrumente*, Rudolf A. Lang Verlag, Pössneck, 1949.

13 Margarida Archinard, *L'apport genevois à l'hygrométrie*, Musée d'Histoire des Sciences de Genève, Geneva, 1980.

14 Roland Wernecke and Jan Wernecke, *Industrial Moisture and Humidity Measurement – A Practical Guide*, Wiley-VCH, Weinheim, Germany, 2014, sec. 4.3.7.

2.2 *Psychrometers (Wet-and-Dry Bulb Hygrometers)*

A psychrometer was made from two identical, matched thermometers (either alcohol or mercury) set side by side, a few centimetres apart. One of them had a dry bulb that measured the ambient temperature T_{db}. The bulb of the other was surrounded with a moist cotton wick, which due to the evaporation of water, displayed a lower temperature T_{wb}. In 1825, Ernst Ferdinand August of Berlin named the device a psychrometer. From then on, the wet-and-dry-bulb hygrometer has been commonly called August's psychrometer, even though he was not the inventor.[15]

Reproducible results could only be obtained with this type of hygrometer if the speed of the air blowing on the wet bulb was higher than 2 to 3 metres/second. To get this velocity, the static August psychrometer was soon replaced by the sling psychrometer, in which the pair of thermometers were whirled overhead by the hand holding them and the Assmann aspiration psychrometer, which had a mechanical fan to draw air over the wet bulb. The sling psychrometer was used mainly onboard ships.

Relative humidity RH was calculated from the two temperature readings and barometric pressure reading by means of semi-empirical formulas, tables, charts, or mechanical devices.[16] The accuracy of RH was limited to a few percent. A complete measurement, including the calculations, took the better part of a quarter of an hour. Psychrometric measurements required skilled personnel. Today, their main use remains in meteorology.[17]

2.3 *Condensation (Dew Point) Hygrometers*

From the measurement of the *dew point temperature* (T_{DP}) at which a slowly cooled mirror was tarnished by water droplets condensing on its surface, along with the *ambient temperature* (T_{amb}), it was possible to calculate the absolute humidity as well as the relative one, according to relationships based on both theoretical and experimental foundations. The mirror-condensation (dew point) hygrometer has for a long time been the most accurate one in common use, mainly for the calibration of other devices in laboratory conditions. The cooling was generally provided by the (smelly!) evaporation of ether. Remote,

15 Middleton, *op. cit.* (n. 10), pp. 125–126.

16 To calculate the relative humidity, it is customary to apply the formula proposed by A. Sprung:

"$RH \approx 100\,[p_{sw} - p_{atm} \cdot A \cdot (T_{db} - T_{wb})] \,/\, p_{sd}$"

where p_{sw} and p_{sd} are the saturation pressures, respectively at the wet bulb temperature T_{wb} and at the dry bulb temperature T_{db}, both calculated by the Magnus / Tetens formula (see Note 9). A is an instrument parameter to be experimentally determined. For standard, well-built instruments, $A = 0.66 \cdot 10^{-3} /\,^{o}C$.

17 Wernecke and Wernecke, *op. cit.* (n. 14), sec. 4.3.3.

telescopic viewing of the thermometer was necessary to avoid the disturbing influence of body temperature and the moist operator's breath. A factor limiting its accuracy was the difficulty in correctly measuring a surface temperature with a bulky liquid thermometer.

Measuring humidity with a dew point hygrometer was a time consuming operation, reserved for skilled operators. The rather laborious numerical calculations, based on the Magnus / Tetens formula, were much simplifed by the use of a graph, such as that shown in *Figure 2.1*. Nowadays, dew point hygrometers have been transformed into easy-to-use, but rather expensive electronic recording instruments, allowing quasi-continuous, remote, and very precise measurements.[18]

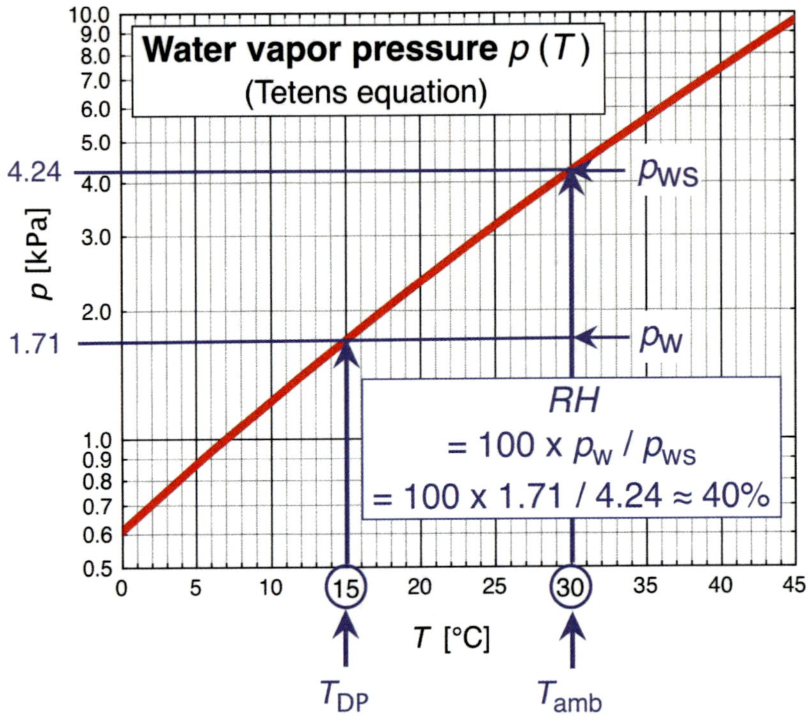

FIGURE 2.1 Example of the graphical determination of the relative humidity *RH* from the ambient and dew point temperature measurements, using the Tetens equation

18 Wernecke and Wernecke, *op. cit.* (n. 14), sec. 4.3.5.

3 Heinrich Greinacher and His Diffusion Hygrometer

Heinrich Greinacher was born in 1880 in St Gall (Switzerland). After studying physics at Zurich, Geneva, and Berlin, he became in 1907 *Privatdozent*, then professor at the University of Zurich where he conducted pioneering studies of the magnetron. He moved to Bern in 1924, where he remained professor at the university until his retirement in 1952. During his long career at Bern, he worked mainly on electrotechnology, radioactivity, and instrumentation for nuclear particle detection (such as spark counters). He died in 1974. His biographer describes him as a very reserved person, preferring to live in isolation, but dutifully and meticulously carrying out his professional tasks.[19]

Greinacher is remembered above all for his voltage doubler (1914), an AC-to-DC converter, followed in 1920 by his cascade voltage multiplier, an AC-to-DC converter producing a continuous output voltage many times larger than the input one. The Greinacher multiplier could also deliver the very high voltages used to test electrotechnical equipment, producing spectacular lightning-like discharges. Swiss and German patents were granted in 1920. The cascade multiplier was later independently rediscovered by John Cockroft and Ernest Walton at the Cavendish Laboratory. They used it in 1932 to build a 700 kV particle accelerator. Consequently, the Greinacher cascade voltage multiplier became known as the Cockroft-Walton cascade voltage multiplier.

3.1 *Greinacher's Double Diffusion Hygrometer*

Heinrich Greinacher was also the Director of the Meteorological Observatory at Bern. At age 64, near the end of his active life – and 70 years after Dufour – he took a sudden interest in instruments to measure the atmosphere's moisture content. His first paper, published in 1944, showed a laboratory prototype of his double diffusion hygrometer.[20] It was followed in 1946 by a more professionally built model (*Fig. 2.2*).[21] It had two identical porous plates placed at the rear, protected from draughts. Light oil manometers measured the pressures.

A schematic drawing (*Fig. 2.3*) explains how it works. In vessel A, the evaporating water increased the pressure relatively to the external atmospheric

19 Heinz Balmer, "Heinrich Greinacher (1880–1974)", *Verhandlungen der Schweizerischen Naturforschenden Gesellschaft* 154 (1974), pp. 239–251.

20 H. Greinacher, "Ein neuer Feuchtigkeitsmesser: das Diffusionshygrometer", *Helvetica Physica Acta* 17 (1944), pp. 437–454.

21 H. Greinacher, "Das Diffusionshygrometer", *Schweizerische Bauzeitung* 127, no. 5 (1946), pp. 52–53; H. Greinacher, "Ein neuer Feuchtigkeitsmesser: das Diffusionshygrometer", *Schweizerische technische Zeitschrift* 13 (1947), pp. 189–192.

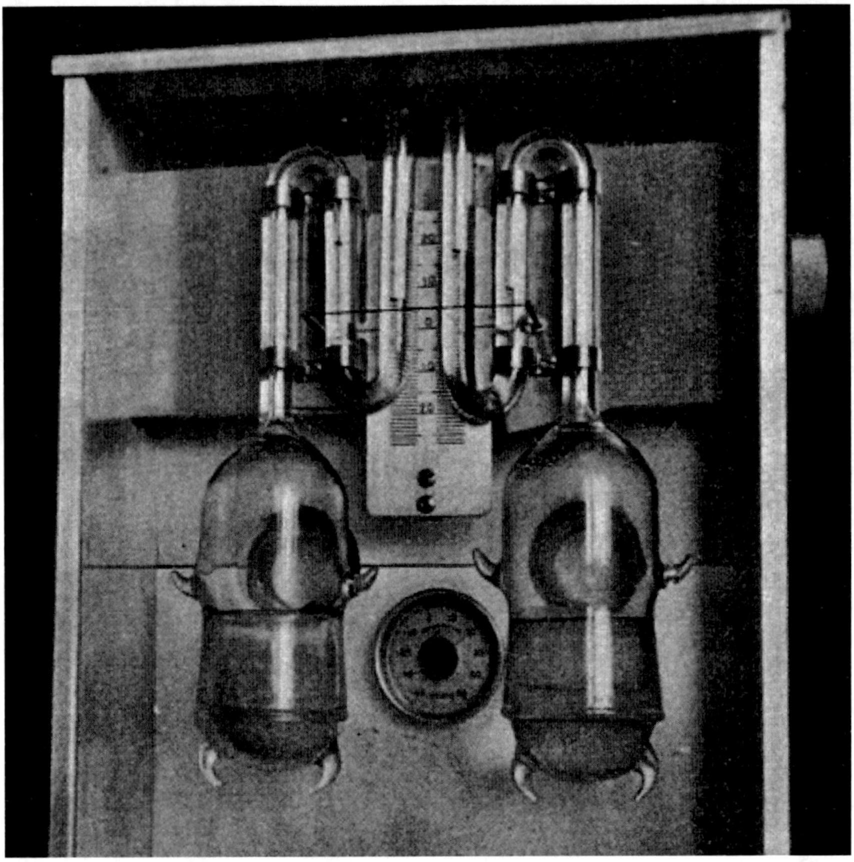

FIGURE 2.2 Heinrich Greinacher's double diffusion hygrometer. The tube on the left contains the desiccant (silica gel); that on the right, distilled water. The graduated scale can be moved up and down. *Schweizerische Bauzeitung* (1946), fig. 2

pressure. In vessel B, due to the absorption of the atmospheric moisture by the desiccant, the pressure decreased, in agreement with the observations of Dufour. Then, a simple calculation gave the value of the relative humidity directly.[22] The result did not depend on the ambient temperature, nor on the

22 $RH = p \,/\, p_s = \Delta p_B \,/\, (\Delta p_A + \Delta p_B)$. The following theory was proposed by Greinacher, "Ein neuer Feuchtigkeitsmesser", *cit.* (n. 20):

> "In the vessel *A* (partly filled with distilled water), the excess-pressure is $\Delta p_A = K(p_s - p)$, where K ($0 < K < 1$) is an instrument constant, depending on the properties of the porous plate.
> Putting the device in dry atmosphere ($p = 0$) allows the measurement of K: $K = \Delta p_A \,/\, p_s$.
> In the vessel *B* (containing an ideal desiccant) the under-presssure is $\Delta p_B = K \cdot p$.
> Thus $\Delta p_A + \Delta p_B = K \cdot p_s$ and $RH = p \,/\, p_s = \Delta p_B \,/\, (\Delta p_A + \Delta p_B)$. Q.E.D.
> The sensitivity, and thus the accuracy, increases linearly with *K*."

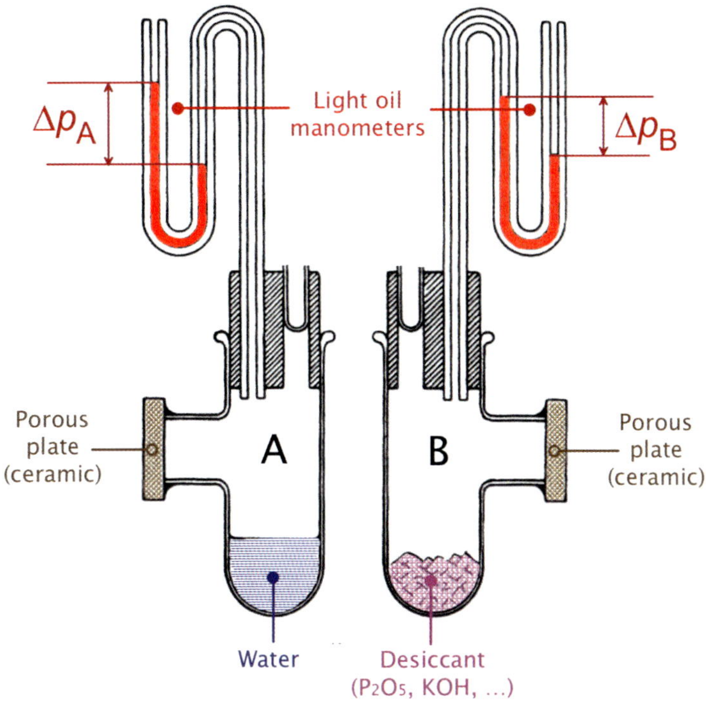

FIGURE 2.3 Schematic drawing of the double diffusion hygrometer. After a long enough time in moist atmosphere ($RH > 0$, partial water pressure $p < p_S$) at ambient temperature, the pressure in vessel *A* becomes greater than the atmospheric pressure (over-pressure Δp_A) and the one in vessel *B* smaller (under-pressure Δp_B), in agreement with the observations of Dufour. Adapted from Greinacher, *Helvetica Physica Acta* (1944), fig. 2

properties of the porous plate. The small thermometer in the same box was actually not required when measuring the relative humidity.

Greinacher got two patents for his invention: a Swiss one (CH 252382) and a British one (GB 612716), both published in 1948. In his papers, Greinacher cited the works of Louis Dufour, as well as the theoretical considerations of Kundt and briefly Schidlowsky, but did not remark on Jouveneau.

3.2 *Greinacher's Simplified Diffusion Hygrometer*

In 1949, without giving reasons for it, Greinacher made a drastic simplification of his double diffusion hygrometer (*Fig. 2.4*), keeping only the vessel *A* with its water filling.[23] Why? My personal supposition is the following: While

23 H. Greinacher, "Das Diffusionshygrometer", *Schweizer Archiv für angewandte Wissenschaft und Technik* 15 (1949), pp. 353–355.

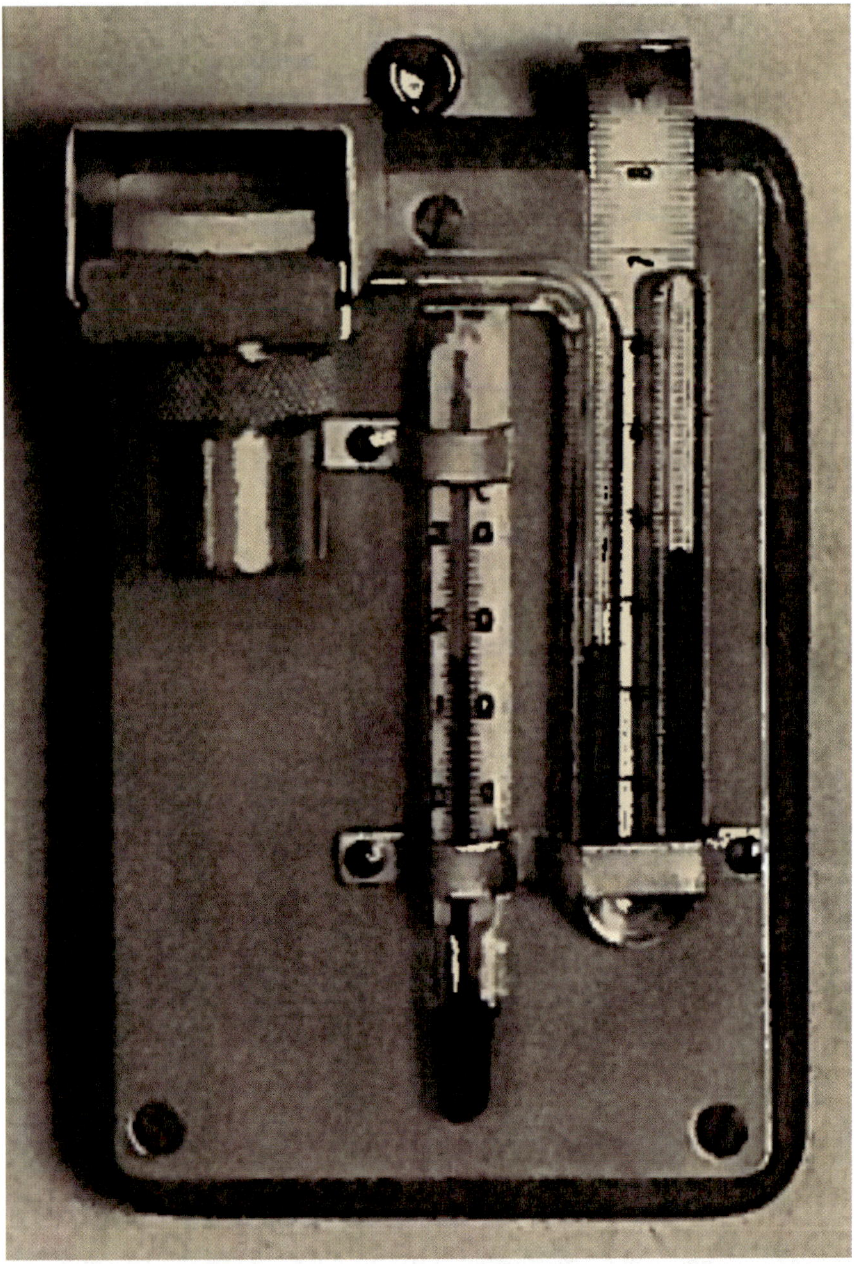

FIGURE 2.4 Greinacher's simplified diffusion hygrometer. *Chimia* (1953), fig. 3

a water refill could be done by anybody, replacing the desiccant when it had become water saturated required a laboratory environment. The desiccant was a powder, and the most effective ones were irritating, corrosive, toxic chemical substances (e.g., P_2O_5, KOH). The measured relative humidity *RH* value now depended critically on the measured temperature T_{amb} and on the properties of the single porous plate, defined by an *a priori* unknown instrument parameter *K*. A numerical table, established for a given value of *K*, could provide the value of *RH* as a function of the measured temperature T_{amb} and of the manometer reading Δp_A.

A more convenient way, proposed by Greinacher, was to read *RH* from a graph (*Fig. 2.5*). The higher the temperature, the more sensitively- and accurately-measured value of *RH* became. From this diagram and from the known value of p_s (T), a rather low value of K (≈ 0.12) could be inferred, limiting the sensitivity and accuracy. The effect of the atmospheric pressure (and thus of altitude) was luckily small and not published by Greinacher before 1953.[24] By its temperature range (5° C to 32 °C), the hygrometer was limited to indoor use.

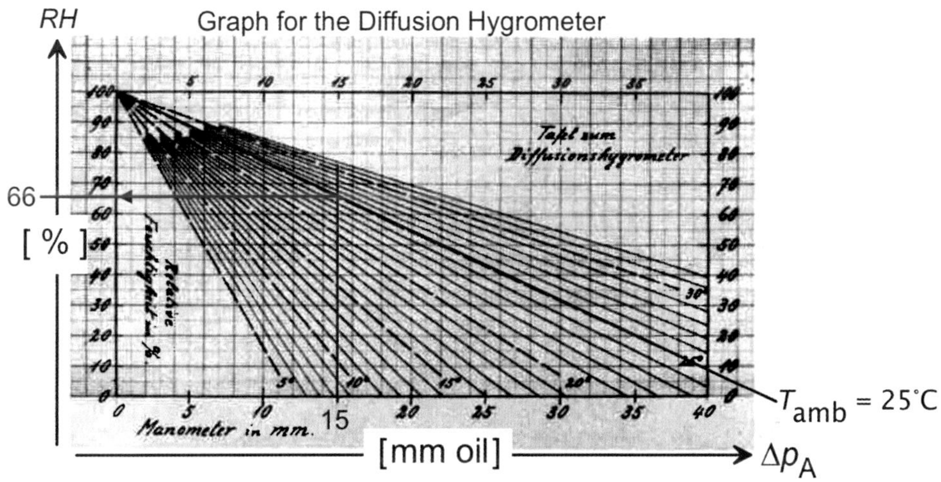

FIGURE 2.5 Example of the graphical determination of the relative humidity from the manometer readout and the temperature. Adapted from Greinacher, *Schweizer Archiv für angewandte Wissenschaft und Technik* (1949), fig. 2

24 H. Greinacher, "Das Diffusionshygrometer, ein neuer Feuchtigkeitsmesser", *Chimia* 7 (January 1953), pp. 11–16.

FIGURE 2.6 Commercial diffusion hygrometers: (a) TESA (height 195 mm); and (b) Rüeger (height 215 mm)
UNIL-EPFL COLLECTION OF SCIENTIFIC INSTRUMENTS, LAUSANNE, 603.0027 AND 603.0510 RESPECTIVELY. PHOTOS BY J.-F. LOUDE, 2017

3.3 *Commercial Diffusion Hygrometers*

In the late 1950s, diffusion hygrometers were produced by two young firms in Lausanne. Rüeger and TESA. An example by each maker is found in the collections of the Physics Museum at Lausanne, both in good condition (*Fig. 2.6a, b*).[25] In different ways, each implemented mechanically the graphical method (see *Fig. 2.5*) proposed by Greinacher.

Rüeger (now Rüeger SA) was founded in 1942 at Lausanne by Ernst A. Rüeger, specialising in the manufacture of bimetallic thermometers. Its first model of a diffusion hygrometer looked very much like the one (*Fig. 2.4*) made in Bern.[26] It was succeeded in 1958 by another "Hygro-Therm" model (*Fig. 2.6b*), which was more convenient to use.[27] Following an idea of Greinacher, a rotating arm and

25 See n. 2.
26 Greinacher, *Chimia*, *cit.* (n. 24), fig. 4.
27 Rüeger, *Gebrauchsanweisung/Mode d'emploi pour l'hygromètre à diffusion "Hygro-Therm", type HR*, Lausanne, 1958, 2 pp.

a sideways-moving manometer assembly offered a relatively easy method for directly reading *RH*.[28] For the necessary periodic water refill, a rather delicate manipulation was needed.

Our Rüeger hygrometer was tested in 2017, more than fifty years after its fabrication, by comparison with a psychrometer and a modern capacitive hygrometer (TESTO 608-H2), accurate to ±2%. The values of *RH*, measured from the beginning of August to the end of October, were systematically much too high.

TESA was founded in 1941 at Renens near Lausanne as Téléphonie S.A., initially making telephone parts. Later, it specialised in precision length-measurement instruments for mechanical workshops. It is still active today.

The TESA model (*Fig. 2.6a*) was designed to be screwed in a strictly vertical position to a solid wall, and was much more convenient to use than the one by Rüeger. One simply read the temperature and the value of *RH* on the rotating acrylic glass drum driven by a bimetallic spiral. The internal construction (*Fig. 2.7*) was original. Periodic water refills (every two to three months, according to the maker) were conveniently made from the top through a hole in the porous plate. A detailed instruction manual was produced with the hygrometer.[29] In the UK, this hygrometer was also available from Rototherm.

Our TESA hygrometer was also tested in 2017. The measured values of *RH* were in excellent agreement (±3%) with the ones given by the capacitive hygrometer and a psychrometer.

Both models required a permanent installation mounted on a wall (particularly the TESA hygrometer, with its suspended mechanical system). The instruments had to be protected from draughts, solar rays, and heat sources. Both used better porous plates ($K = 0.25 \pm 0.01$) than Greinacher's prototypes, thus greatly improving their sensitivity and accuracy.

4 Diffusion Hygrometers: Reception and Conclusions

4.1 *Reception*

The experimental results obtained with the diffusion hygrometers provided data useful for a comparison of diffusion theories (e.g., the Knudsen Law and

28 H. Greinacher, "Diffusionshygrometer mit direkter Ablesung", *Schweizer Archiv für angewandte Wissenschaft und Technik* (June 1954), pp. 198–200.

29 TESA S.A., *Instructions de mise en marche, mode d'emploi, … de l'hygromètre* TESA (n. d.), 8 pp.

FIGURE 2.7 Internal view of the TESA hygrometer
PHOTO BY HJ. BÜHLMANN, 2017

others) by Spencer-Gregory and Rourke.[30] The TESA / Rototherm model was mentioned in a journal for possible use in museum collections.[31] Otherwise, diffusion hygrometers can rarely be found in books and papers about hygrometry, and then they appear only as a curiosity. One mention by Wexler and Brombacher reported the diffusion hygrometer to be extremely sensitive to rapid temperature changes.[32] Spencer-Gregory and Rourke offered a five-page description and included it in a short chapter investigating porosity in solids.[33] In 1964, Lück underscored the difficulty in finding satisfactory materials for the porous membranes, and the consequent lack of technical interest in this type of hygrometer.[34] The last known reference known to me of the diffusion hygrometer appeared in a paper by Johnson, published in 1978; he pointed to their sluggish response and their very high sensitivity to quick changes in temperature, wind, dust and solar radiation.[35]

Concerning the commercial propagation of the diffusion hygrometers, I have not been able to find examples in other collections of scientific instruments, apart from the two specimens at Lausanne.[36]

4.2 *Conclusions*

The diffusion hygrometer was certainly a working instrument. But was it, as promoted by Greinacher, actually better than other contemporary hygrometers? As this paper has shown, it was not.

- It was adapted only to indoor use, due to the limitation of its temperature range.
- The claims of high accuracy were not backed up by serious calibration data.
- The influence of atmospheric pressure, and therefore of the altitude, on the measurements, was not well studied (an effect that is particularly important in a mountainous country like Switzerland).
- The long-term stability of the instrument's measurements was questionable.

---

30 H. Spencer-Gregory and E. Rourke, "A Theoretical Basis for the Diffusion Hygrometer", *Philosophical Magazine* 38 (1947), pp. 573–580.

31 Raymonde Frin (ed.), "Climatology and Conservation in Museums", *Museum* 13, no. 4 (1960), p. 254, fig. 20, and p. 260.

32 Wexler and Brombacher, *op. cit.* (n. 11), p. 1.

33 H. Spencer-Gregory and E. Rourke, *Hygrometry*, Crosby Lockwood & Son, Ltd., London, 1957, pp. 36–40, and Chapter 8.

34 Winfried Lück, *Feuchtigkeit*, R. Oldenbourg, München-Wien, 1964, pp. 147–148.

35 Dale L. Johnson, *Instruments for Measuring the Amount of Moisture in Air*, NASA Technical Memorandum TM-78190, August 1978, p. 21.

36 Recently (August 2022) a private collector informed me that he had inherited a TESA hygrometer from his grandfather, who had a clock shop in Zurich in the 1960s. Its serial number is 1267, higher than the one (724) on our model.

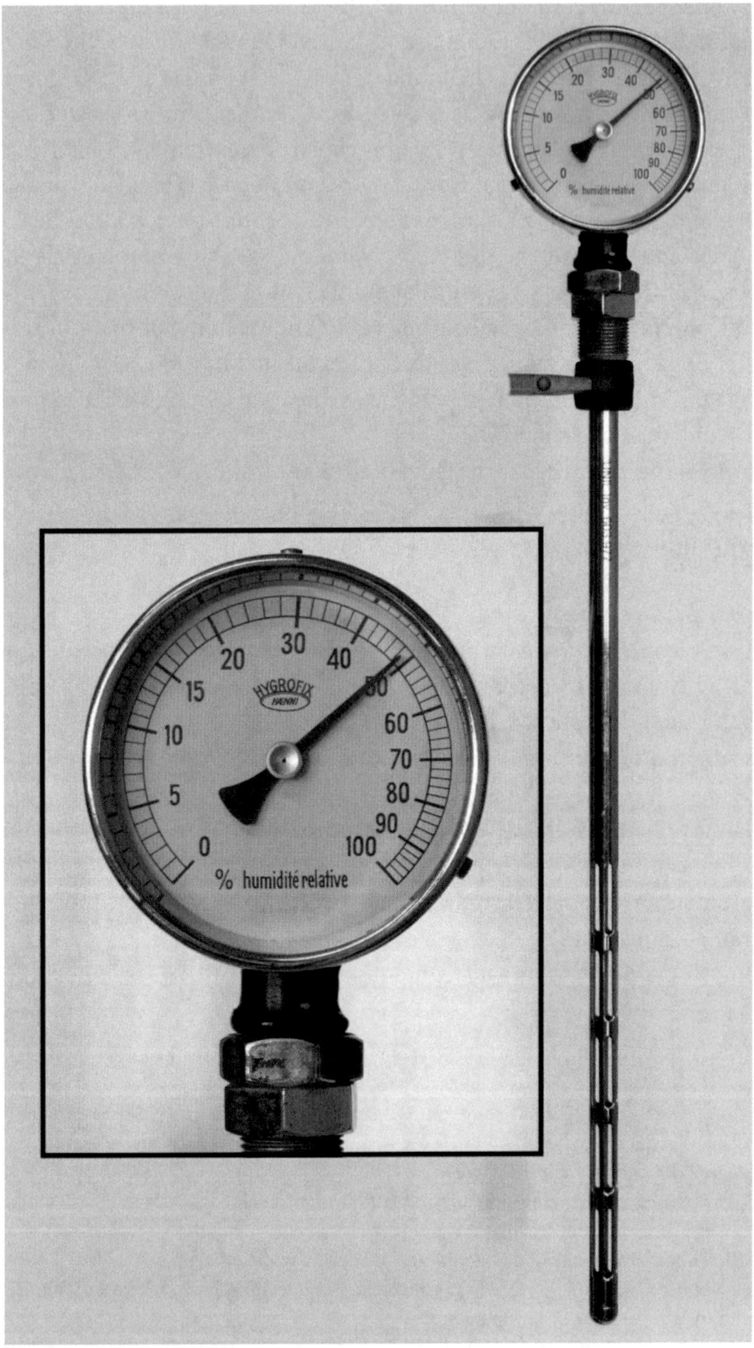

FIGURE 2.8 Hair hygrometer "Hygrofix Haenni" (dial diameter 100 mm)
UNIL-EPFL COLLECTION OF SCIENTIFIC INSTRUMENTS, LAUSANNE, 603.0977. PHOTO BY J.-F. LOUDE, 2017

- There were other serious drawbacks:
  - The calibration was based on the *a priori* unknown properties of a porous plate.
  - The device was not portable, because any contact of the porous plate with the manometer oil had to be avoided at all costs.
  - No recording version of the instrument was produced.
- Concerning manufacturing of the instrument, the availability of porous plates with reproducible properties would have certainly been a problem.

The diffusion hygrometer by Rüeger or TESA was not better than a good hair hygrometer of the day (such as one by Haenni near Bern, *Fig. 2.8*).[37] The hair hygrometer was much easier to use, offered an accuracy of ±3% at 20 °C and was working in a much larger range of temperatures.

After 1965, advances in electrical and electronic technology made new types of hygrometers available (e.g., LiCl dew cells, capacitive and resistive hygrometers). Initially difficult to operate and reserved for laboratory use, the dew point hygrometer, at last, was transformed into an easy-to-use, high-precision (±0.1%) instrument. It benefitted from much more convenient and odourless cooling elements (electrical Peltier elements), electrical temperature measurement, LED illumination of the mirror, and photo-electrical detection of the apparition of the dew. The instrument offered immediate and almost continuous evaluation and display of the measurements, made possible by micro-controllers. The related dew cells became the preferred instrument of the meteorologists, who nonetheless continued to use psychrometers when no power supply was available. But the cheap hair hygrometer was and still is widely produced, sold and used for non-critical applications, especially indoors.

Acknowledgements

I am very grateful to the late Hansjörg Bühlmann for returning both hygrometers to working condition and testing them. Surprisingly, the firm Rüeger was able to find the missing "Mode d'emploi" of their Hygro-Thermo of 1958. I have also to thank the École Polytechnique Fédérale de Lausanne, through my laboratory, the Laboratoire de Physique des Hautes Energies, for its continued support. Last but not least, I would like to express my very great appeciation to Sara Schechner for her valuable comments during the reviewing of this paper.

37 Haenni & Cie A. G., *Handbuch über Manometer, Thermometer, Hygrometer*, catalog 14, group 0801–0807, Jegenstorf, Switzerland, 1953.

CHAPTER 3

"Extremely Poor": an Equatorial Mount That Failed

Johan Kärnfelt

On the evening of 7 December 1816, the staff at Stockholm Observatory were readying the equipment for a night of observations.[1] The transit clock had been adjusted earlier in the day and, by 9 PM, director Simon Anders Cronstrand, cartographer Carl Peter Hällström, and janitor and factotum Gustaf Wiklund had adjusted the instruments and were ready to get to work. First off was a lunar occultation, in which the Moon was to occult the star ϰ Geminorum, a standard observation in the age of celestial mechanics.[2] Cronstrand used the observatory's workhorse, a 10-foot refractor made by Dollond in London that had been in service for many years of observations; Wiklund was at the new 7-foot Herschel reflector; and Hällström used his private telescope from the yard outside the observatory's main hall, where his colleagues were located. Cronstrand noted the immersion as $9^h\ 12^m\ 27^s$, and later the emersion to $10^h\ 8^m\ 49^s$ or 50^s. Wiklund, probably due to his lesser experience, marked both events as two seconds later, and Hällström, who was counting the ticks of another clock, obtained similar figures. Following the successful observation, there was just enough time to take some meridian transits before the Stockholm sky clouded over.[3]

At first glance, there is nothing extraordinary about this scene – an astronomer and his assistants working at their instruments, struggling with timekeeping and battling the weather. The odd one out is the Herschel telescope; representing a major investment and with a construction process that dragged over two decades, this seems to have been the only time it was used under the stars. In an article on the fragility of scientific instruments, historian Simon Schaffer states that "faults are defaults, yet instruments perform", but this was

1 On the history of the observatory, see Steven Haywood Yaskell, "From Research Institution to Astronomical Museum: A History of the Stockholm Observatory", *Journal of Astronomical History and Heritage* 11 (2008), pp. 146–156.

2 According to www.calsky.com, the occultation appeared between $19^h\ 55.3^m$ and $20^h\ 49.6^m$ CET. Cronstrand and his colleagues noted the event in sidereal time, thus the time difference.

3 Simon Anders Cronstrand, Journal, 5 October 1813–25 May 1819, Simon Anders Cronstrand Papers: Stockholm, Royal Swedish Academy of Sciences (RSAS), Center for History of Science. All archival sources in this paper are from this repository. Translations from Swedish by the author.

 | DOI:10.1163/9789004689107_005

not the case with the Herschel telescope.[4] Regardless of the successful observation of ϰ Geminorum, it was quite simply a total failure.

In this article, I will attempt to answer the question of why the Herschel telescope failed, and discuss the consequences of that failure. The narrative starts with the cumbersome process that resulted in a finished telescope, moves on to the technical aspects of its failure, and finishes with a sketch of the telescope's more recent history.

1 The Mirror

Made out of brass, steel, and mahogany, the telescope carries at its heart a speculum mirror that was moulded and polished by William Herschel. With an aperture of 16 cm and a focal length of 210 cm, the Herschel telescope of the Royal Swedish Academy of Sciences remains an impressive piece of equipment (*Fig. 3.1*).[5] However, on closer inspection, the mount seems somewhat out of place, not in keeping with the style of the tube or that of the other Herschel telescopes that are preserved in collections around the world (compare *Fig. 3.5*).[6] The archives tell us that this was a later addition, a matter to which we will return.

The optics – primary and secondary mirrors, lenses for two finders, and five eyepieces – were delivered in December 1790. Together, they constitute the vital heart of a reflecting telescope, a type of instrument suggested by Isaac Newton a hundred years earlier.[7] The speculum mirror (*Fig. 3.2*) was made from an alloy of copper and tin, which allowed the metal to be polished to a high reflectivity by a technique Herschel had mastered. The bill exceeded 31 pounds sterling, a considerable sum at the time.[8]

---

4 Simon Schaffer, "Easily Cracked: Scientific Instruments in States of Disrepair", *Isis* 102 (2011), pp. 706–717: 707.

5 On the history of the Academy, see Johan Kärnfelt, Karl Grandin, and Solveig Jülich (eds.), *Knowledge in Motion: The Royal Swedish Academy of Sciences and the Making of Modern Society*, Makadam, Gothenburg, 2018. See also Sten Lindroth, *Kungl. Vetenskapsakademiens historia, 1739–1818*, Almqvist and Wicksell, Stockholm, 1967.

6 Andreas Maurer, "A Compendium of All Known William Herschel Telescopes", *Journal of the Antique Telescope Society*, 14 (Winter 1998), pp. 4–15.

7 C. A. Ronan, "The origins of the reflecting telescope", *Journal of the British Astronomical Association* 101 (1991), pp. 335–342.

8 Receipt, no. 217, 1790. On Herschel's telescope business, see Michael Hoskin, "Herschel Telescopes for Sale", *Journal for the History of Astronomy* 46 (2015), pp. 291–296. His 7-foot telescope, similar in size to the Swedish Academy's instrument, is discussed in Jim Bennett, "'On the Power of Penetrating into Space': The Telescopes of William Herschel", *Journal for the History of Astronomy* 7 (1976), pp. 75–108: 76–77.

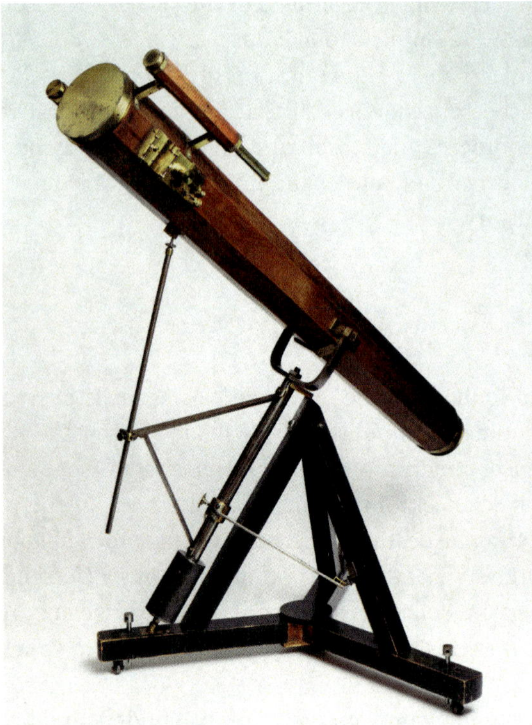

FIGURE 3.1
The Herschel telescope of the Royal Swedish Academy of Sciences (RSAS), on its present mount (RSAS/Per Myrehed)

FIGURE 3.2 The 16 cm speculum mirror made by William Herschel and delivered to the Royal Swedish Academy of Sciences in 1790 (RSAS/Per Myrehed)

The circumstances of the acquisition are shrouded in mystery. It is not clear who approached Herschel with the order, or even what the Academy wanted to achieve with the new instrument. The archives of Academy astronomer Henric Nicander and secretary Johan Carl Wilcke offer no indications that either of them took an interest in the matter. Wilcke might even have been opposed to the acquisition. In a letter to botanist Olof Swartz, he sarcastically wrote that when the telescope was ready, "one believes one will be able to count stamen and pistils on the Moon".[9] However, there was one member of the Academy who was known to be interested in the Herschel telescopes, Baron Peter Niclas von Gedda.

Von Gedda was a judge at the Administrative Court of Sweden, but he was also trained in the arts and had a keen interest in mechanics. He was elected as an Academy member in 1786 and, at about the same time, gave a speech at one of its meetings on recent developments in science, during which he highlighted Herschel's discoveries of nebulae and briefly discussed "the Herschelian and Newtonian tubes and their precedence".[10] True to his views, in 1792 he presented the Academy with a reflector that he had built around a speculum mirror produced by a friend. According to von Gedda, this was the first working reflecting telescope made in Sweden.[11] A few years later, he tried without success to cast his own speculum mirror for a 9-foot telescope.[12] With this background, and in the light of his later involvement with the Herschel telescope, it is reasonable to assume that it was von Gedda who, as a newly elected member and with the blessing of the Academy, initiated the deal with Herschel.

In February 1788, there was a message from England stating that the mirror was ready to be shipped. The matter was discussed by the *Inspectura Aerarii*, the Treasury Inspection, which granted payment. At the same meeting, von Gedda offered to "take care of the necessary gantry" – i.e. to ensure the mirror was mounted in a telescope.[13] The good Baron appears to have taken his time. As stated above, the optics arrived in 1790, but the minutes do not note that von Gedda had made arrangements with instrument maker Carl Apelquist before 1795.[14]

Even in its early years, the Academy realised the importance of collaborating with skilled instrument makers, with the highly regarded Daniel Ekström even having his workshop in the basement of the Academy Observatory, until

9 Quoted from Lindroth, *op. cit.* (n. 5), p. 318.
10 Peter Niclas von Gedda to Johan Carl Wilcke, n.d. [letter a], Johan Carl Wilcke Papers.
11 Minutes RSAS, 13 July 1792, Section 1.
12 Peter Niclas von Gedda to Johan Carl Wilcke, n.d. [letter b], Johan Carl Wilcke Papers.
13 Minutes Inspectura Aerarii, 29 February 1788, Section 3.
14 Minutes RSAS, 9 September 1795, Section 5.

his premature death in 1755.[15] Apelquist belonged to the next generation of instrument makers.[16] Around 1775, he started a two-year apprenticeship with Jesse Ramsden in London and, on returning to Sweden, he received a grant from the Royal College of Commerce to travel back to London as an apprentice to William Fraser.[17] This was supposed to last four years, but Apelquist returned a year later, in the spring 1778, claiming to be "fully trained".[18] During this period, he also managed to buy some of the tools from James Short's estate.[19] Back in Sweden, he first set up a workshop in the city of Karlskrona, making instruments for the navy, before, some years later, moving his business to Stockholm. In 1784, he applied to the King in Council for *privilegium exclusivum* for making certain instruments, which the Academy, as one of several referral bodies, endorsed, but with the addendum that some of his apparatus were copies rather than original constructions.[20] Despite the verdict, the king granted the privilege in 1785 and, over time, the workshop became a thriving enterprise, specialising in making industrial equipment like machine tools and jackscrews.[21] In 1794, astronomer Henric Nicander suggested his election to the Academy, but the vote was not in favour.[22]

2 The Mount

After the announcement that Apelquist was to build the telescope, the minutes went quiet. Many years passed, and the Academy seems to have forgotten about the arrangement. However, in 1804, a decision was made to conduct an inventory of the equipment at the Academy Observatory. When the inspectors reported back, it was clear that everything was "in good order and nothing is missing, except the mirrors to a Heshelian [sic] tube which are being stored with Major

15 Olov Amelin, *Medaljens baksida: Instrumentmakaren Daniel Ekström och hans efterföljare i 1700-talets Sverige*, Institutionen för idé- och lärdomshistoria, Uppsala, 1999.

16 On Apelquist, see Lindroth, *op. cit.* (n. 5), pp. 807–808; Amelin, *op. cit.* (n. 15), pp. 147–150. See: Anita McConnell, *Jesse Ramsden (1735–1800): London's Leading Scientific Instrument Maker*, Ashgate, London, 2007, p. 62.

17 Minutes RSAS, 4 June 1777, Section 4.

18 Minutes RSAS, 21 March 1778, Section 1.

19 A. D. Morrison-Low, *Making Scientific Instruments in the Industrial Revolution*, Ashgate, Aldershot, 2007, p. 141.

20 Minutes RSAS, 15 December 1784, Section 3.

21 Minutes RSAS, 21 September 1785, Section 3; and Lindroth, *op. cit.* (n. 5), p. 808.

22 Minutes RSAS, 19 November 1794, Section 2.

Apelquist".[23] The forgotten telescope was a bit embarrassing, and the Academy immediately decided on an on-site inspection of Apelquist's workshop.

A few weeks later, the envoys reported back. The mount was "half-finished" and, even worse, "so clumsy and roughly made, that it surely needed the assistance of 4 persons to move back and forth across the floor".[24] Not only this, but Apelquist requested 650 *riksdaler* for the work so far, which, it was noted, was more than it would cost to buy a similar and complete telescope straight from English instrument makers. Maybe the Academy should let Apelquist keep the mirror, and use the funds to buy a working instrument from London? Undecided, the Academy forwarded the matter to the office of Treasury Inspection for further discussion.[25] A few months later, the Inspection office decided to demand the return of the mirror and to have someone else mount it.[26] However, during a second visit to Apelquist's workshop, things became more complicated. It turned out that the clumsy mount design was not Apelquist's idea, but had come from a figure in a book supplied by Academy secretary Wilcke. Apparently, Apelquist was not to blame. To save the day, the Academy offered either to buy the half-finished mount or to let him complete it in a reasonable time.[27] Apelquist met them halfway; he would now accept 300 *riksdaler* for the mount, or he could have it finished in two years.[28]

Once again, the deal ground to a halt and was not resumed until 1811. According to a long account in the minutes from the Academy's December meeting, this was due to a lack of funding, which had prevented the Academy from closing the deal. Now a settlement was made, and the still half-finished mount was bought for 333 *riksdaler* from Apelquist. Instrument maker and microscope specialist Peter Hellström was to finish the telescope. "Thus the R. Academy could hope to soon see a vital scientific need met, even if the execution was mightily expensive for the R. Academy, but its honour demanded it."[29] It was indeed expensive, but financial support came from an unexpected source. At the same meeting, there was an announcement that the King was to provide 500 *riksdaler* from his personal accounts. With funding secured, Hellström could get to work. He delivered the finished instrument

23 Minutes RSAS, 4 April 1804, Section 7.
24 Minutes RSAS, 5 September 1804, Section 8.
25 Minutes RSAS, 16 June 1804, Section 8.
26 Minutes RSAS, 5 September 1804, Section 3.
27 Minutes RSAS, 24 October 1804, Section 14.
28 Minutes RSAS, 14 November 1804, Section 3.
29 Minutes RSAS, 4 December 1811, Section 2.

to the Academy Observatory in December of 1812, twenty-two years after the Herschel mirror's arrival in Stockholm.[30]

3 Equatorial Mounts

The first mount has not been preserved. Fortunately, the archive holds a lengthy technical description of the original Herschel telescope, written by Academy astronomer Simon Anders Cronstrand in 1813, just a few months after the instrument was finished.[31] The report also contains a drawing by fortification officer Jacob Johan Tavaststjerna (*Fig. 3.3*).

A quick look at the image reveals that this is an equatorial mount, where the polar axis aligns with the Earth's axis. Before further analysing the image, we must take a short detour into the history of these mounts.

We have known that the Earth revolves on its axis since the days of Copernicus. We cannot sense this rather swift movement, but it can be inferred through the motion of the celestial sphere, where stars seem to drift from east to west. The effect is more striking when seen through the eyepiece of a telescope, as celestial bodies move quite rapidly through the field of view. Thus, to observe an object for an extended time, you must move the telescope accordingly. Since the stars' trajectories appear to follow curved paths, great circles perpendicular to the Earth's axis, nudging the telescope somewhat to the west is not enough, it must also be either raised or lowered slightly. On traditional altazimuth mounts, this is transformed into adjustments on two different axes: up-down and east-west. However, by aligning the telescope's polar axis with the Earth's axis, the altitude axis becomes redundant. To track an object with an equatorial mount, you only have to turn the telescope on one axis (i.e., in right ascension).

The equatorial mount has a history that precedes the telescope. The armillary sphere, known since antiquity, and the torquetum, used in medieval Europe since at least the thirteenth century, were both constructed around an equatorial polar axis.[32] Later the basic principle was implemented in telescope mounts, and we find an early example in Christoph Scheiner's helioscope used

30 Hellström invoiced 250 *riksdaler* to finish the telescope. Receipts, 1812, no. 56. Along with the 31 pounds to Herschel and the 333 *riksdaler* to Apelquist, this makes it a very expensive telescope. An estimated price, using historical converters, is about €17,000.

31 Simon Anders Cronstrand, "Ritning och Beskrifning på det genom Hans Majt Konungens frikostighet på Observatorium i Stockholm uppställda Catadioptriska Telescop", MS liggfol.

32 See Giancarlo Truffa, "Rings, Armillae and Other Spherical Instruments", *Proceedings of 2018 IEEE International Workshop on Metrology for Archaeology and Cultural Heritage*

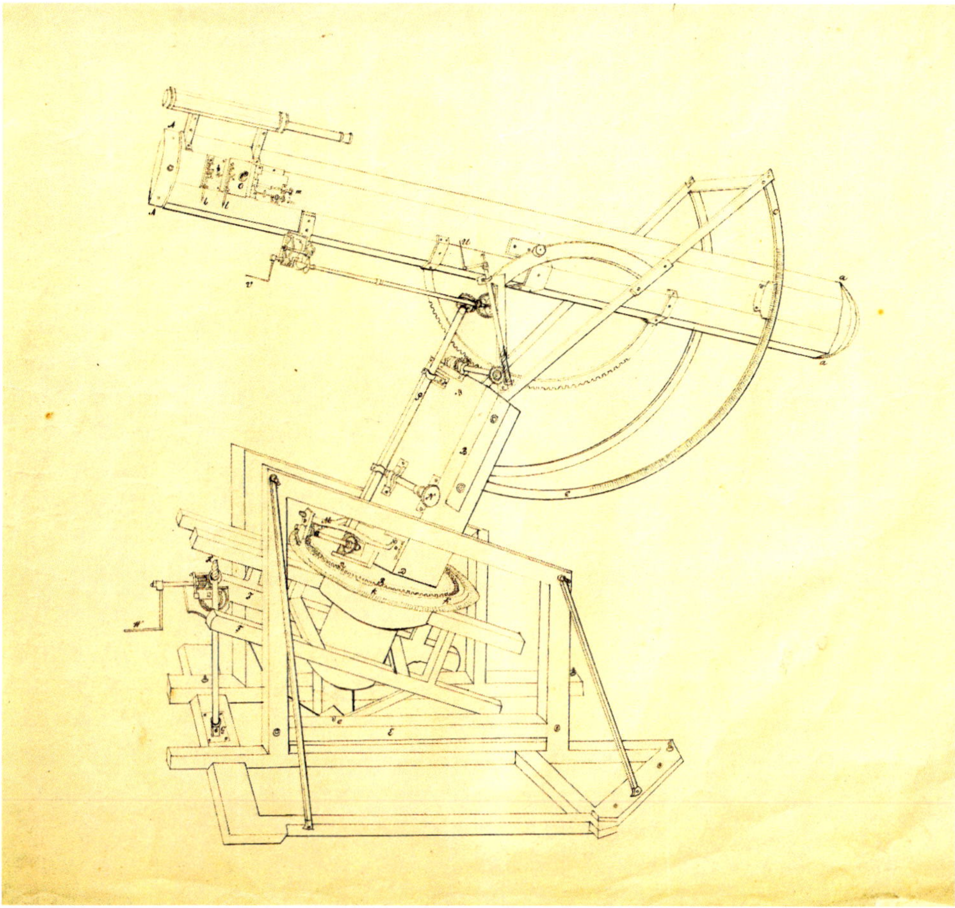

FIGURE 3.3 The Herschel telescope in its original mount, illustrated in a drawing by Jacob Johan Tavaststjerna, 1813 (RSAS)

in the 1620s. Despite attempts by other instrument makers, mounting a telescope on an equatorial did not become popular until the end of the next century, with weight distribution being a major technical challenge. Due to the tilt of the polar axis, one had to compensate for the load on the north side of the mount, which increased with the size of the telescope. This could be done either by supporting the top of the polar axis, sometimes referred to as the English equatorial mount, or by counterweights, later known as the German equatorial. In the 1670s, an early version of the English equatorial was built by John Flamsteed at Greenwich Observatory, to support a 7-foot sextant. The technique

(2018), pp. 544–548; Jaroslaw Wlodarczyk, "The Torquetum (or Turketum): Was It an Observing Instrument", *Journal for the History of Astronomy* 53 (2022), pp. 266–299.

was later implemented for telescopes and tested by many instrument makers, among them Jesse Ramsden who, in 1791, completed an equatorial mount that supported a refractor with a 4-inch aperture for George Shuckburgh.

Meanwhile, the development of the German equatorial followed a similar trajectory. James Short successfully mounted a small reflector on an equatorial in 1749, and many were to follow in his footsteps: Edward Nairne made a portable equatorial in 1771; Nicolas Noël mounted a large Cassegrain telescope with a 61 cm mirror on an equatorial in 1772 (it did not work properly, but it was still the largest telescope in the world until William Herschel entered the stage). If they worked, these early equatorial mounts, based on the torquetum, could only support relatively small instruments, and it was not until German instrument-maker Joseph Fraunhofer perfected the use of counterweights for the Dorpat refractor, finished in 1824, that the equatorial technique set a new standard (and gave the solution its name).[33]

That said, let us return to the Herschel telescope. When Apelquist was commissioned to build the telescope, the equatorial mount was still very much under development. However, rumours had reached Stockholm, and the original mount for the Academy's reflector, as depicted in *Figure 3.3*, had some striking similarities with instruments built abroad. It is possible that Apelquist picked up some design features during his apprenticeships in London, but we also know that Academy secretary Wilcke had approached Apelquist with a drawing in a book, "to which the Mount in question has been, for the most part, adapted".[34] Which specific image Apelquist was shown is impossible to say, but candidates might be found in a couple of articles published in the *Philosophical Transactions of the Royal Society*, a publication that was avidly studied by Academy fellows. For example, James Short had published an article on the aforementioned equatorial in 1749, and in 1771 Edward Nairne had given an account of a "portable Observatory". Both articles contained detailed drawings of the telescopes.[35] Even if Apelquist was struggling with a much larger

33 The history of the telescope is discussed at length in Henry C. King, *The History of the Telescope*, Charles Griffin & Company Limited, London, 1955. See also Maurice Daumas, *Scientific Instruments of the Seventeenth and Eighteenth Centuries and Their Makers*, B. T. Batsford, London, 1972, chap. 3.

34 Minutes RSAS, 24 October 1804, Section 14.

35 James Short, "A letter from Mr. James Short, F.R.S. to the President, with the description and uses of an equatorial telescope", *Philosophical Transactions of the Royal Society* 46 (1750), pp. 241–246; and Edward Nairne, "Description and use of a new constructed equatorial telescope or portable observatory, made by Mr. Edward Nairne, London", *Philosophical Transactions of the Royal Society* 61 (1771), pp. 107–113. A description of Nairne's telescope, including the image from *Philosophical Transactions of the Royal Society*, was also published in the second edition of Jérôme de Lalande's *Astronomie*, Paris, 1771, pp. 667–668,

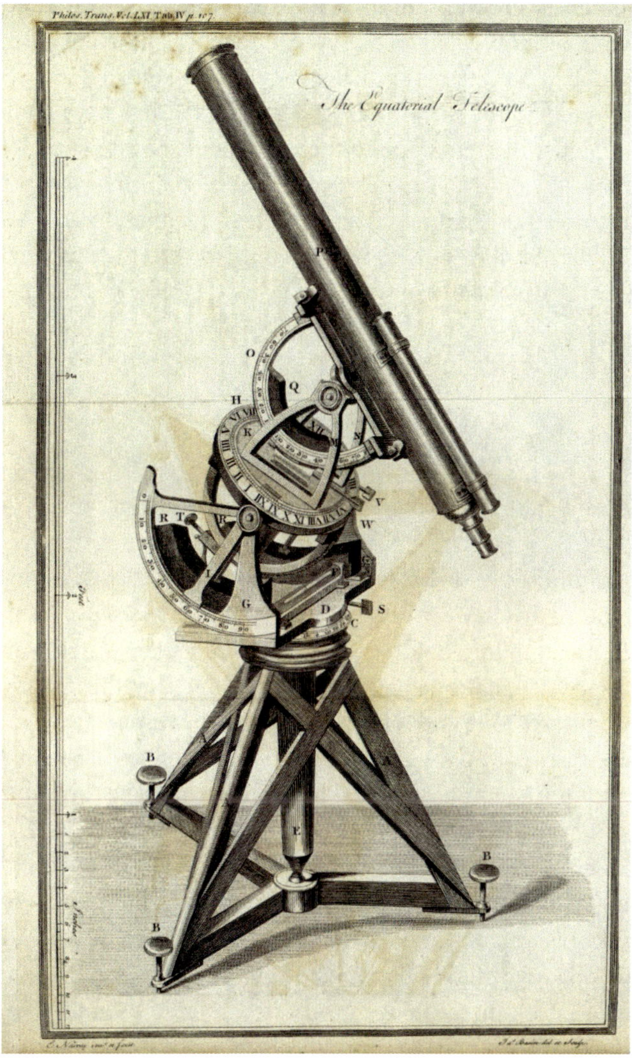

FIGURE 3.4 Edward Nairne's 1771 portable equatorial telescope, as depicted in Edward Nairne, "XIV. Description and Use of a New Constructed Equatorial Telescope or Portable Observatory, Made by Mr Edward Nairne", *Philosophical Transactions of the Royal Society* 61 (1771), pp. 107–113

telescope, his equatorial mount was similar to Nairne's construction (*Fig. 3.4*). So for the sake of argument, let us assume that this was the actual source of inspiration.

---

pl. 2. A copy was available at the Academy, and this might be a second candidate to the "book" that Wilcke showed Apelquist.

Nairne's telescope moved around four axes. Starting from the bottom, the telescope turned around pillar E. This oriented the telescope, so the polar axis pointed towards the north. Next, the tilt of the mount was adjusted at G to make the polar axis parallel to the Earth's axis. Both these settings were made only once in order to adjust the telescope to the latitude of the site at which observations were made. The last two axes came into play during observations and were used to set the right ascension (H) and declination (O). We can also note that weight distribution was not a major concern for this small telescope, but Nairne still attached two counterweights (Q and R) to balance the telescope.

4 Apelquist's Solution

Apelquist's mount looked quite different from Nairne's solution, but the design elements were similar (*Fig. 3.3*).[36] Whereas Nairne's tube was supported by a sturdy brass construction, almost the full weight of Apelquist's tube rested on a toothed semicircle mounted below the tube. However, the tube was also partly supported by the brass scaffolding holding the large semicircles, which were used to set the telescope's declination angle. Bearing in mind that this mahogany tube was over two metres long and included a metal mirror, a focuser, and various fittings, there was a lot of weight pressing on the cog and the small shaft that supported it. Another problem was that the tube would be extremely sensitive to vibration because it rested on only a thin brass semicircle. This alone would have made it very difficult to use.

The tube and the declination arms are mounted on top of the polar axis. Apelquist did not make his axis out of brass as Nairne did, but out of wood. It was hexagonal at the top and cylindrical at the bottom, with the rounded end projecting 15 cm down into a metal housing, where it rotated around a peg inserted into the bottom of the polar axis. To aid this rotation, metal bearings were mounted inside the metal container that supported the axis. There was a cogwheel on top of the container, with a divided scale for taking readings of the right ascension. Through additional linkages, this cogwheel was connected to a crank close to the eyepiece, which let the operator move the telescope in right ascension. As there were no counterweights, the inclination of the polar axis put a large load on the north side which, in turn, put great strain on the peg the axis rode upon, as well as on the roller bearings hidden in the metal housing. Even if the bearings could hold the weight, the resulting friction would have made it very difficult to turn the telescope around its polar axis.

36 My account draws entirely on Cronstrand, *op. cit.* (n. 31).

The metal housing carrying the polar axis was mounted inside a wooden cradle that could pivot on a shaft at its lower end. To align the telescope's polar axis with the Earth's axis, a crank was used to adjust the cradle's inclination. Of all the features of Apelquist's design, this is the most curious. The load on the bottom shaft was enormous, not to mention the load on the crank mechanism, which was the only part stopping the telescope from tipping over. Without additional support, the mechanism was likely to snap. One way of overcoming this would have been to build the mount with a tilt that was fixed to the latitude of Stockholm Observatory, but apparently the telescope mount was designed to be adjustable for different geographical locations. This suggests that Apelquist was following in the footsteps of his English colleagues, who designated their equatorials as "portable".

In summary, there were at least three problems that rendered Apelquist's telescope more or less useless: the support for the tube was not sturdy enough and would have made it extremely sensitive to vibrations; without counterweights, the friction created by the tilt would have made turning the telescope in right ascension very difficult; and the design of the support for the polar axis put a lot of stress on the altitude mechanism, making it likely to break. In addition, the mount was quite bulky and probably weighed well over 100 kg.

5 Lessons Learned

As stated in the introduction, the Herschel telescope seems to have been used only once. No indications of other use survive in the archives or in any published material. However, we know that observatory director Simon Anders Cronstrand was unhappy with the new instrument and, even if he did not mention any of the above problems in his technical report, his views can be inferred from a later document.

In 1819, he raised the issue of new instrumentation needed for the observatory at the Academy, later submitting a list of things he believed necessary for upholding international scientific standards.[37] The Academy reviewed his suggestions, and a document from 1819, signed by Uppsala astronomers Zacharias Nordmark, Jöns Svanberg, and Johannes Bredman, outlined the basic requirements. The reviewers suggested investments in two pendulum clocks, a transit instrument, and a Ramsden circle, all instruments necessary for the positional astronomy then in vogue at the observatory. At the end of the document, however, they also discussed the "Astronomical Observations, particularly

37 Minutes Inspectura Aerarii, 19 May 1819, Section 6.

characteristic of Schröter and Herschel", i.e. lunar and nebular observations. Even if they did not endorse this kind of research, they pointed out that:

> It cannot pass unnoticed, that such Observations above all demand strong Reflectors, of which the Academy, sad to say, does not own any that are suitable. We mention this with real sadness, with regard to the expense the Academy sustained in acquiring the so-called *Herschelian* 7-foot reflector. Unfortunately, that is extremely poor, much worse than the larger *Gregorian* reflector the Academy used to own, or some of its better Achromatic tubes.[38]

"Extremely poor". The reviewers did not qualify the statement, but it is quite clear that the Herschel telescope did not work as intended. Since lunar and nebular astronomy, in the reviewers' opinion, were of lesser importance to scientific progress, and since funds were limited, and there were great needs elsewhere, they believed attempts to remedy this situation would be ill-advised. However, they continued:

> This is reminiscent of the necessity, in the event of the commissioning of any Instrument, to turn not only to the best workshops, but also to correspondence with practical Astronomers active in the field, who with zeal embrace every opportunity to benefit their science. This will ensure proper examination of the commissioned instrument before it is supplied, so that it may, in fact, possess all the perfection at which one previously took a cheap chance.[39]

Given that the scientific outcome of the Herschel telescope was nil, this seems reasonable. In fact, the lesson learned may well have been the most important outcome of the business with the Herschel telescope.

6 A New Mount

The fate of the Herschel telescope during the 1820s and 1830s is unclear, but one might wonder why the astronomers did not try to salvage the mirror and the tube. Installed in a much simpler altazimuth wooden mount, as in some

---

38 Zacharias Nordmark, Jöns Svanberg, and Johannes Bredman, Med oinskränkt delande …, 5 October 1819, appendix.

39 *Ibid.*

FIGURE 3.5 Telescope built by William Herschel for William Watson at Bath in 1783–1785
SCIENCE MUSEUM, 1876–1000 PT 1. © THE BOARD OF TRUSTEES OF THE SCIENCE MUSEUM, LONDON

of the telescopes William Herschel built himself (*Fig. 3.5*), the telescope could have enabled the lunar and nebular astronomy neglected by the observatory. The answer might have been a combination of professional embarrassment and the price tag. Scrapping the mount would have been an admission that the Academy had thrown money down the drain, and that they had failed miserably in their management of the affair. Maybe it was better to forget about the whole thing, and let the telescope collect dust in storage at the observatory.

A remounting of the telescope would not happen before the observatory got a new director.

In the late 1810s, Cronstrand focused on geodetical research, participating in expeditions across Sweden and leaving observatory-bound astronomy to his assistants. Until his retirement in 1837, activities at the observatory were more or less at a standstill. It was not until the reign of his successor, Nils Haqvin Selander, that the Herschel telescope was dusted off. In 1842, Selander decided to scrap "the failed old parallactic mount" and let "Conducteur Gernandt" make a new one.[40] The resulting mount still carries the telescope today.

The reasons for Selander's involvement with the Herschel telescope are obscure – no explanations can be found in the inspection reports or minutes that can shed any light on his endeavour. In addition, the choice of instrument maker was odd. At the time, the Academy had two official instrument makers, but Edvard Gernandt was not one of them. He started his career as a workshop manager at the Technological Institute, the forerunner of the present-day Royal Institute of Technology, but left the institute in 1839 to set up his own workshop, later specialising in clock movements for churches and belfries.[41] We do not know why Selander turned to Gernandt, but a possible reason was that Selander wanted to test the newcomer on the reasonably simple construction that became the new mount (*Fig. 3.1*). There are no preserved invoices from Gernandt, but he received a small grant from the Academy in 1843. According to the minutes of the Treasury Inspection office, he had earned "the Academy's encouragement", and the grant might have been compensation for his work on the mount.[42]

The new mount, a German equatorial, was a fairly neat, much simpler construction than Apelquist's. The technical solution might have been supplied by Selander, but it could equally well have been suggested by Gernandt himself, and as a part of the challenge. Anyway, Gernandt opted for a polar axis with a fixed tilt adapted for the latitude of Stockholm Observatory, and balanced the tube with an adjustable counterweight. In so doing he reduced the bulk of the mounting, solved the problems Apelquist had struggled with, and got the telescope in working order.

Once on its new mount, the Herschel telescope embarked on a new career, not as an instrument of science but as a demonstration instrument and historical showpiece. It was on display at the Observatory, and sometimes pulled out to let visitors get a glimpse of the stars.

---

40 The Observatory inspection report 1842, 12 April 1843.

41 J. Lundin, "C. F. E. Gernandts levnadshistoria", *Tid-Skrift* 5 (2013), pp. 5–8.

42 Minutes Inspectura Aerarii, 3 February 1843, Section 5.

However, the handling of the instrument through the years took its toll, especially after 1931 when it was put on display at the new observatory in Saltsjöbaden. Per Olof Lindblad, who had grown up at the observatory and later became its director, remembered how the damaged telescope had been stored away (with the mirror removed and on display in the lecture hall).[43] Once again collecting dust, the telescope did not resurface until 1962, when it was a showpiece in the Nordic Museum's exhibition, "The Heritage from Newton and Linnaeus".

That part of the telescope's saga started with celebrations occasioned by the Royal Society of London's tercentenary in 1960.[44] King Gustav VI Adolf and Academy secretary Erik Rudberg represented Sweden at the festivities. After the celebrations, the Academy was offered the opportunity to borrow and display a poster exhibition on British history of science, which had been prepared for the tercentenary. The offer was accepted, and the exhibition went on display at Uppsala University the next year. The occasion later inspired an Academy member, librarian William Odelberg, to put together a more ambitious exhibition that narrated, as the subtitle said, the "Scientific Links between England and Sweden in Bygone Times". In the winter of 1961–1962, instruments, rare books, manuscripts, and paintings were gathered, some of them on loan from England, and a lengthy exhibition catalogue was produced. To live up to the demands of public scrutiny, the Herschel telescope was renovated. The mahogany tube, which had cracked during storage, was mended, cleaned, oiled, and waxed; the brass fittings were refurbished and varnished; the mount, made from steel and covered in rust, was sanded down and oiled. Eyepieces and finders also got a makeover.[45]

When the exhibition opened in 1962, the telescope was on display (*Fig. 3.6*). Needless to say, the telescope's messy history was not outlined. Instead the instrument was contextualised by Lemuel Francis Abbot's famous portrait of William Herschel, on loan from England. The catalogue remarked that Herschel made at least 430 speculum mirrors and sold them to customers all over Europe:

> However, it is depressing to experience the slight extent to which these instruments have contributed to the development of astronomy, as most of them did not end up with scientists but with dilettantes, especially

43 Per Olof Lindblad to Johan Kärnfelt, 11 August 2019, author's private archive.

44 Wilhelm Odelberg, *Lärdom för livet*, Atlantis, Stockholm, 1999, pp. 359–361.

45 The renovation process is outlined in J. Lindgren to RSAS, 8 June 1962, Wilhelm Odelberg Papers.

FIGURE 3.6 On exhibition in Stockholm in 1962, the Herschel telescope protrudes from behind the white rectangular pedestals. A portrait of William Herschel is on the wall to the left
PHOTO FROM THE EXHIBITION, *HERITAGE FROM NEWTON AND LINNAEUS*, RSAS

> considering the enormous importance of William Herschel's own research into the emergence and development of modern astronomy. For example, it was with an instrument of this type and size that he discovered the planet Uranus on March 13, 1781.[46]

There are several things worth noticing in this passage. We might disregard the fact that Herschel himself was an amateur, or dilettante in the language of the time. Moreover, as we have seen, things did not always go well when a Herschel mirror ended up in the hands of the "scientists" such as those who populated the Academy of Sciences. But even such a Swedish technical failure might become a success when serving as a stand-in for telling a significant

46 *The Heritage from Newton and Linnæus*, in *Bidrag till Kungl. Svenska vetenskapsakademiens historia* 2, Royal Swedish Academy of Sciences, Stockholm, 1962, p. 103.

historical account such as the discovery of Uranus. Once linked to Herschel's most famous discovery, the telescope became part of a grand narrative of magnificent discoveries and scientific progress, the antithesis of the trajectory that it had actually followed.

7 Conclusions

The Herschel telescope of the Royal Swedish Academy of Sciences failed, and did so for a number of reasons. First, Apelquist's equatorial mount was, from a technical point of view, more or less a disaster. Its problems are so obvious that one can wonder why Apelquist, who was an experienced instrument maker, did not consult his client about them. Maybe he was too embarrassed to raise the question, or the lack of a suitable counterpart at the Academy made him hesitate. In his defence, it must be underscored that his instructions were vague, to say the least, and that the design of the equatorial he was trying to reproduce was still very much under development, a technology that had not yet come to complete fruition amongst his foreign colleagues.

However, the failure must also be attributed to the Academy's handling of the affair. As noted by the Uppsala astronomers in their review of Cronstrand's suggestion for new instrumentation, there was a lack of management of the telescope's commissioning. The Academy did not seem to have had a clear idea of what they wanted to achieve with the new instrument, they had even fussier ideas about its construction, and above all, they did not supervise the production process.

Lastly, and this might be the most striking fault, the Academy seems to have accepted the failure without taking any further steps to remedy the situation. Despite it being an expensive and potentially very useful instrument, they simply let it fade away. As noted above, they could have employed a carpenter to make a simple altazimuth mount, allowing the telescope to fulfil at least some of its potential. If they had done so, this might have sparked an early interest in nebular astronomy, which was gaining traction elsewhere, but did not enter the research agenda of Stockholm Observatory until the early 1900s.

Since the 1960s, the Herschel telescope has upheld its role as an exhibition showpiece and has been on display many times.[47] However, it has always been contextualised in relation to discovery and progress, or to the celestial wonders. However, given the results of this paper, there seems to be a very different

47 The telescope's later history is discussed in Johan Kärnfelt, "Vetenskapsakademiens Herschel-teleskop: En instrumentbiografi", *Slagmark* 37 no. 81 (2020), pp. 133–154.

exhibition story to tell, a story about trials and errors, about the slow development of technology, and about the many failures that are also necessary for the progress of science.

Acknowledgements

My sincere gratitude goes to the staff at the Center for the History of Science, Royal Swedish Academy of Sciences, to the community of the Science Instrument Commission, especially Richard Kremer, to the editors of this volume, Sara Schechner and Sofia Talas, and to my language editor Clare Barnes.

CHAPTER 4

A Telescope's Lack of Fortune: the Newall Telescope

Panagiotis Lazos and George N. Vlahakis

In the present paper we shall study a telescope commissioned by the English wealthy entrepreneur, Robert Stirling Newall (1812–1889). Completed in 1869, "Newall's telescope" was both a technological masterpiece and a failure. A careful examination of contemporary references to the telescope did not find a single remark on any inadequacy in its construction. On the contrary, both the original optical and mechanical parts were so good that even today original parts of the instrument are still working. Nevertheless, we consider it a failure because it was scarcely used during the first twenty-five years of its existence. The first paper using observations done with this telescope was only published in 1892.

What made it a failure was the way in which Newall managed the telescope. We will show that he did not hire an expert to use the instrument during the time he was engaged with his duties as an industrialist. The site of the instrument near Newcastle was also problematic from the start. Air pollution from adjacent factories and humidity from the nearby River Tyne dramatically reduced the quality of the observations. Once it was moved to Cambridge, the instrument flourished in the hands of Newall's son, Hugh Frank Newall (1857–1944).[1] He published twenty papers of observations in eighteen years, a fact that proves that the instrument itself was not a failure, but its use by Newall senior.

1 Robert Stirling Newall: Engineer, Industrialist, and Astronomer

Robert Stirling Newall is an exemplary figure of nineteenth-century amateur astronomy in Britain. He was born in Dundee, Scotland in 1812. During his youth, he worked in London under Robert McCalmont experimenting on methods to produce steam rapidly. In 1838 at the age of twenty-six and inspired by a letter from a friend, L. D. B. Gordon who was visiting the mines in Germany, Newall designed a wire rope machine, which produced a four-stranded rope with four wires to each strand. A year later, Newall formed a partnership with Gordon

1 H. A. Brück, "Obituary: Hugh Frank Newall", *The Observatory* 65 (1944), pp. 195–199.

 | DOI:10.1163/9789004689107_006

and Charles Liddell to set up the R. S. Newall and Company in Dundee to make wire rope. Another factory was founded in Gateshead, England, and the firm saw even better days because of its role in developing improved submarine cables and manufacturing the first submarine cable successfully laid between Dover and Calais in 1851.[2] Other important dates in the history of Newall's enterprises were 1853 when he invented the "brake-drum", 1855 when his company submerged a number of cables in the Mediterranean, and 1868 when he supervised the line connecting Ringkjobing in Denmark with Newbiggen, Northumberland.[3]

From these enterprises, Newall had the money to be engaged "above the average" with his other great love, astronomy. Inspired by his friend Hugh Lee Pattinson, FRS,[4] whom had acquired a 7.5-inch equatorial telescope from the instrument maker Thomas Cooke of York in 1851 and set up an observatory at his residence, Scots House, near Gateshead, Newall started negotiations to buy a telescope from Cooke as early as 1860. Having common interests both in business and astronomy, the friends' connection was also strengthened in 1849 when Newall married Pattinson's daughter Mary, and together they had four sons and one daughter.

Newall's contributions to industry and astronomy of the United Kingdom were acknowledged with his election as a fellow of the Royal Astronomical Society in 1864, of the Royal Society in 1875, and as a member of the Institute of Mechanical Engineers in 1879. He was decorated with the order of the Rose of Brazil in 1872, and an honorary degree was conferred upon him in 1887 by the University of Durham. Newall died at his home, Ferndene, at Gateshead-on-Tyne on 21 April 1889.

2 Thomas Cooke and the Construction of the Telescope

Thomas Cooke was born in 1807 in Allerthope, Yorkshire. He came from a very poor family, but despite many difficulties, he managed to study mathematics and optics. At the age of twenty-two, he made his first achromatic lens. This success prompted him to start a small optical company in York. Cooke's

2 Robert Stirling Newall, *Facts and Observations Relating to the Invention of the Submarine Cable, and to the Manufacture and Laying of the First Cable Between Dover and Calais in 1851*, E. & F. N. Spon, London, 1882.

3 A. M. Clerke, "Newall, Robert Stirling", in Leslie Stephen and Sidney Lee (eds.), *Dictionary of National Biography*, London, Smith, Elder, & Co., 1885–1901, 66 vols: v. 40, pp. 309–310.

4 Bernard Barham Woodward, "Pattinson, Hugh Lee", in *Dictionary of National Biography. op. cit.* (n. 3), v. 44, pp. 56–57.

methodical and accurate work made him famous as a maker of optical materials like lenses. An additional fact that helped his company flourish was the railway connection between York and London.[5]

He was the first to establish an equatorial telescope factory in Britain. He specialised in the manufacture of high-quality, medium-sized telescopes, but also manufactured other optical instruments like microscopes. In the first years of his factory, he built twelve telescopes with a diameter of 5 to 8 inches and nine telescopes with a diameter of 8 to 10 inches. He was soon followed in this work by his two sons, Frederick and Thomas.

It is not a surprise that Cooke accepted Newall's invitation to build the largest telescope in the world. This would be a real challenge for him personally and his company in general. Reportedly, Newall's inspiration for this enterprise was the availability of two glass disks, 2 inches thick and 26 inches in diameter, produced by the glass firm Chance Brothers of Birmingham and shown at the 1862 London International Exhibition (which took place in South Kensington).[6] However, D. W. Dewhirst has claimed that the discussions between Newall and Cooke for the creation of the telescope started at least two years earlier.[7]

The final agreement was ready in September 1863.[8] Newall purchased both glass disks for £1000,[9] and they were sent to Cooke's workshop.[10] Cooke promised to build the telescope in twelve months – in part to secure Newall's order and in part because he had a high confidence in his skills – but difficulties appeared during the project. According to Dewhirst, Cooke "had to work four optical surfaces whose area was almost double that of any previous lens ... and design and make the huge equatorial mounting in his small factory".[11] Cooke used some ingenious techniques to improve the optical quality of the

5 A. D. Morrison-Low, "Spirit of Place: Some Geographical Implications of the English Provincial Instrument Trade, 1760–1850", *Bulletin of the Scientific Instrument Society*, no. 53 (1997), pp. 19–24: 23.

6 Norman Lockyer, *Stargazing, Past and Present*, MacMillan, London, 1878, p. 305; "Robert Stirling Newall", *Monthly Notices of the Royal Astronomical Society* 50, n. 4 (1890), pp. 165–167, https://doi.org/10.1093/mnras/50.4.165.

7 D. W. Dewhirst, "The Newall Telescope", *Journal of the British Astronomical Association* 80, no. 6 (1970), pp. 493–495: 493.

8 "Newall", *op. cit.* (n. 6), pp. 165–167. Newall ordered also a 7-inch meridian circle from Cooke.

9 Dewhirst, *op. cit.* (n. 7), p. 493.

10 "[Newall] ... did not hesitate to risk thousands of pounds in one great experiment, the success of which will have a most important bearing upon the astronomy of the future." Lockyer, *op. cit.* (n. 6), p. 302.

11 Dewhirst, *op. cit.* (n. 7), p. 494.

disks. For example, during the grinding process of shaping the lenses, the two glass disks were floated in mercury to protect their exact shapes from wrong movements.[12] This procedure put artisans at risk from the inhalation of mercury vapour. It took 1560 hours of grinding for the disk's surfaces to acquire the necessary curvature, with a simultaneous loss of one inch of thickness.[13] The work on the objective lens was not finished until 1865, when it was presented at the British Astronomical Association's meeting in Newcastle.[14]

Nevertheless, the whole project became long overdue, and the relationship between Newall and Cooke worsened over time. Newall insisted that his telescope should be given first priority by Cooke, and he nagged him. As both a financier and an entrepreneur, he demanded satisfaction for the terms of delivery. Cooke suffered from the pressure that Newall exerted on him. The awkward situation gradually weakened his health, and Cooke died in 1868. The telescope was finally completed by Cooke's sons.

The technical specifications of the new telescope impressed both experts and the public. The aperture of the object glass was 25 inches; the focal length was 29 feet; and the lens weighed 144 lbs. The objective had two elements separated by a layer of air no more than 12 mils (a few tenths of a millimetre) thick. The first element was made of crown glass with a central thickness of 1.85 ± 0.125 inches (4.7 ± 0.3 cm) and refractive index from 1.51065 to 1.52759 depending on the wavelength. The second element was made of flint glass with a central thickness of 0.9 ± 0.08 inches (2.3 ± 0.2 cm) and a refractive index from 1.61243 to 1.64607.[15]

The tube had a cigar shape for minimum weight. It was 32 feet long and had a diameter of 34 inches in the centre but only 27 inches at each end. It was made of steel plates in five lengths. Inside there were five zinc tubes. The distance between them was about an inch, and they served to limit hot air currents that would interact with light travelling down the tube. Sir W. G. Armstrong of Elswick-on-Tyne made the telescope's pedestal.[16] The height of the cast-iron pillar was 29 feet, and its base had a diameter of 5 feet 9 inches. The whole

12 [Charles Pritchard], "Obituary – Thomas Cooke", *Monthly Notices of the Royal Astronomical Society*, 29 (1868), pp. 130–135: 134.

13 Lockyer, *op. cit.* (n. 6), p. 128.

14 E. A. Milne, "Hugh Frank Newall, 1857–1944", *Biographical Memoirs of Fellows of the Royal Society* 4, no. 13 (November 1944), pp. 717–732, https://doi.org/10.1098/rsbm.1944.0018.

15 D. Papathanasoglou and J. Deliyannis, "A Method for the Determination of the Optical Constants of a Large Refractor", *Monthly Notices of the Royal Astronomical Society* 205 (1983), pp. 1097–1102, https://doi.org/10.1093/mnras/205.4.1097.

16 Milne, *op. cit.* (n. 14).

instrument (including the tube, mount, and pier) weighed nearly 9 tons, but the telescope tube could easily be moved with one hand.[17]

There were three finders. Two of them had a diameter of 4 inches and were located on opposite sides of the tube. This way the observer always had one available, no matter the position of the telescope. The third finder was larger with a diameter of 6.5 inches, and it had been installed mainly for finding comets.[18]

At the time of its completion, the Newall telescope was by far the largest refracting telescope in the world, taking the place of the 18.5-inch refractor by Alvan Clark and Sons at the Dearborn Observatory in Chicago.[19] In 1873 its size was surpassed by the 26-inch refractor at the U.S. Naval Observatory, also made by Clark.[20] It must be emphasised that before making the 26-inch telescope, Clark had visited the Newall telescope, accompanied by Simon Newcomb, Asaph Hall, and William Harkness, in order to examine it and probably to learn about the design of its construction.[21]

3 First Stage: the Telescope at Gateshead

The telescope was completed in 1870[22] and installed at Newall's house, Ferndene,[23] near Gateshead, on the River Tyne near Newcastle (*Fig. 4.1*). To help in the construction and erection of the telescope, Newall had recruited the experienced German astronomer Albert Marth (1828–1897) in 1868. Marth had worked in Malta with the English merchant and astronomer William Lassell (1799–1880).[24] It is not clear how long Marth worked at Gateshead, but he left before 1883.

---

17 "The Newall Telescope", *Nature* 1 (17 February 1870), pp. 408–410.

18 "The Newall Telescope", *The Engineer*, 13 May 1870, p. 289.

19 "Newall Telescope", *op. cit.* (n. 18).

20 The discs of glasses were made again by Chance. Lockyer, *op. cit.* (n. 6), p. 119.

21 Lockyer, *op. cit.* (n. 6), p. 309.

22 The year 1871 is often mentioned as the year of completion of the telescope; see for instance Milne, *op. cit.* (n. 14), p. 718. The description of Holiday makes it clear that the telescope was installed and functional in the dome as early as May 1870. However, some definitive additions (e.g., the clockwork mechanism, the micrometer, and the illumination) were installed or completed in 1871. See Henry Holiday, *Reminiscences of My Life*, W. Heinemann, London, 1914, pp. 169–173.

23 For more details about Newall's house see: https://twsitelines.info/SMR/7711.

24 E. B. K., "Obituary Notices: Fellows: Marth, Albert", *Monthly Notices of the Royal Astronomical Society*, 58 (1898), pp. 139–142: 141. The fact that Newall had hired Marth two years

FIGURE 4.1 The Newall telescope at Ferndene without its dome in 1877
COURTESY OF CAMBRIDGE UNIVERSITY, INSTITUTE OF ASTRONOMY LIBRARY

Newall had initially wanted to put the observatory in Madeira and make it available to invited astronomers.[25] This plan explains why Newall designed the dome to be portable. It could be easily dismantled, transported, and re-erected. The reasons why Newall did not carry out this project were first, Cooke's delay and later, Newall's lack of time because of his work.[26] In his own words:

> I have for some years intended to send my 25-inch telescope, when finished, to Madeira, for the sake of a better atmosphere and a more southern latitude; but as the completion of it has been delayed so much longer than I ever anticipated, I have now decided to keep it in England and ... I now think of removing it to the South of England.[27]

Newall never stopped looking for a better place for his unique telescope. When in 1875 there was some discussion in Britain about the formation of a "Physical Observatory", Newall offered to lend the telescope for work in "Physical Astronomy", but the whole idea did not go ahead.[28]

In May 1879, Newall proposed to David Gill (1843–1914), an astronomer at the Royal Observatory at the Cape of Good Hope, to lend the telescope to the Cape Observatory for seven years. The only condition Newall set was to have the telescope returned to him without incurring shipping costs. As Gill was informed of the offer on the day before his departure for South Africa, a committee was assembled to consider Newall's proposal. It included William Spottiswoode (1825–1883), a mathematician, physicist, and expert in light polarisation; William Huggins (1824–1910), an expert in astronomical spectroscopy; James Nasmyth (1808–1890), an engineer and inventor; Warren De La Rue (1815–1889), an expert in astronomical photography; Sir William Siemens (1823–1883) with whom Newall did business in Bulgaria and elsewhere; and Lord James Lindsay (1847–1913), the president of the Royal Astronomical Society. Although the observatory urgently needed a new telescope, the

---

before the actual erection of the telescope proves he thought the project would be finished earlier.

25 Dewhirst, *op. cit.* (n. 7), p. 494.

26 "Newall", *op. cit.* (n. 6).

27 R. S. Newall, "Mr Newall's Great Refractor", *Astronomical Register* 103 (July 1871), p. 167. See also "Flotes and gossip", *Photographic Art Journal*, 1, n. 1 (March 1870), p. 27. The rumours circulated about this telescope demonstrate how much it had impressed the public.

28 "Newall", *op. cit.* (n. 6), p. 166.

committee decided that the total cost of this proposal, estimated at around £3000, was too much for a seven-year loan.[29]

The telescope remained at Ferndene for twenty years where it idled and did not contribute anything much to the exploration of the skies or the progress of astronomy. This complete lack of scientific work with one of the world's largest telescopes is surely some kind of failure even though it is not a technical one.

The causes of the failure were varied. First, Newall was required to be regularly absent for business. As a result, his observations were infrequent and certainly unsystematic. In addition, many nights with suitable weather conditions were spent unexploited. An even more serious problem was that Ferndene was a place completely unsuitable for astronomical observations. The area had low altitude and was located in the north of England where the percentage of clear nights was low.[30] Within a short distance from the observatory, the River Tyne and the sea both increased humidity in the atmosphere. Aside from meteorological factors, there was also intense air pollution. In the nineteenth century, fast industrial growth was accompanied by a rapid increase in northeast coal production and use. Tyneside became a major industrial area, and Gateshead's population increased from 15,000 to 50,000 in a few decades.[31] We now understand what the astronomical author W. F. Denning meant when he wrote, "The large Cooke refractor … has a singularly barren record. Its atmospheric surroundings appear to have rendered it impotent". Newall agreed. In a letter sent to Denning in 1885, Newall wrote, "Atmosphere has an immense deal to do with definition. I have only had one fine night since 1870! I then saw what I have never seen since".[32]

This combination of pleasant surprise and problems with the atmosphere was felt by Henry Holiday, a pre-Raphaelite artist and amateur astronomer,[33]

29 Anon, "Proceedings of Observatories. Royal Observatory, Cape of Good Hope", *Monthly Notices of the Royal Astronomical Society*, 40 (1880), pp. 232–236: 236, doi: 10.1093/mnras/40.4.236. Gill left for Cape in May 1879: see John S. Reid, "David Gill FRS (1843–1914): The Making of a Royal Astronomer", *Journal for the History of Astronomy*, Vol. 49 (1) (2018), pp. 3–26, doi: 10.1177/0021828617751290.

30 The sky is clear less than one night at four or even ten, depending on the season. https://weatherspark.com/y/42225/Average-Weather-in-Gateshead-United-Kingdom-Year-Round#Sections-Precipitation (accessed 9 November 2022).

31 Sitelines – Gateway to the Tyne and Wear's Historic Environment Record, Industrial Period (1750–1950), https://twsitelines.info/industrial-period (accessed on 9 November 2022).

32 William F. Denning, *Telescopic Work for Starlight Evenings*, Taylor, London, 1891, p. 25.

33 Holiday was able to keep "one eye on the telescope and one on the paper" as his friend and astronomer Norman Lockyer used to say.

when he made some observations in 1870–1871.[34] Holiday had met Newall in 1870. He had been drawing the Moon with the help of a 3-inch refractor of his own, but the idea of using the largest telescope of the time fascinated him. His brother-in-law, the Reverend Milville Raven, was an acquaintance of Newall, so Holiday dared to write a letter to Newall asking permission to observe the Moon with the "monster telescope" and draw a crater. Newall responded positively to the request. As soon as the Moon was again in the first quarter, on 6 May 1870, the Newalls warmly welcomed Holiday. Let us see the dome and the telescope through his words:

> The dome was fifty feet in diameter, and this vast metal structure was so beautifully mounted that one person could make it revolve with ease so as to bring the opening opposite the right part of the heavens. There was also a platform on circular rails, which not merely followed any required movement of the telescope round the observatory, but was so constructed that it could ascend and descend in a curved line corresponding with the upward and downward sweep of the eye-piece; and each movement was practicable by one person, who had only to turn a handle [*Fig. 4.2*]. All this I was able to see by daylight, and after dinner Mr. Newall and I went to the observatory to begin work.[35]

The view of the Moon through the eyepiece caused even greater surprise:

> The sight was overwhelming. The brilliancy of the light, and the scale of these rocky mountains regions, which looked as if one could climb about in them, made the prospect so exciting that it was difficult to sit down calmly and begin to draw, much as I wanted to do so.[36]

On that night, Holiday completed the first lunar-crater sketch ever made with the Newall telescope; he depicted the Maurolycus crater (*Fig. 4.3*). Holiday donated a copy to Newall, who had been at the finder and had manually kept the Moon in the telescope's field of view, as the clockwork mechanism had not yet been placed on the instrument.[37] It was reported that Holiday had to use an eyepiece with moderate magnification (60×) because of humidity in the atmosphere.

---

34 There are four such sketches of Moon's craters and of sunspots in the Institute of Astronomy Library of the University of Cambridge. https://www.repository.cam.ac.uk/handle/1810/216784 (accessed on 9 November 2022).

35 Holiday, *op. cit.* (n. 22), p. 170.

36 *Ibid.*

37 Holiday, *op. cit.* (n. 22), pp. 170–171.

FIGURE 4.2 Robert Newall on the observing platform mentioned by Henry Holiday, operates his telescope
COURTESY OF CAMBRIDGE UNIVERSITY, INSTITUTE OF ASTRONOMY LIBRARY

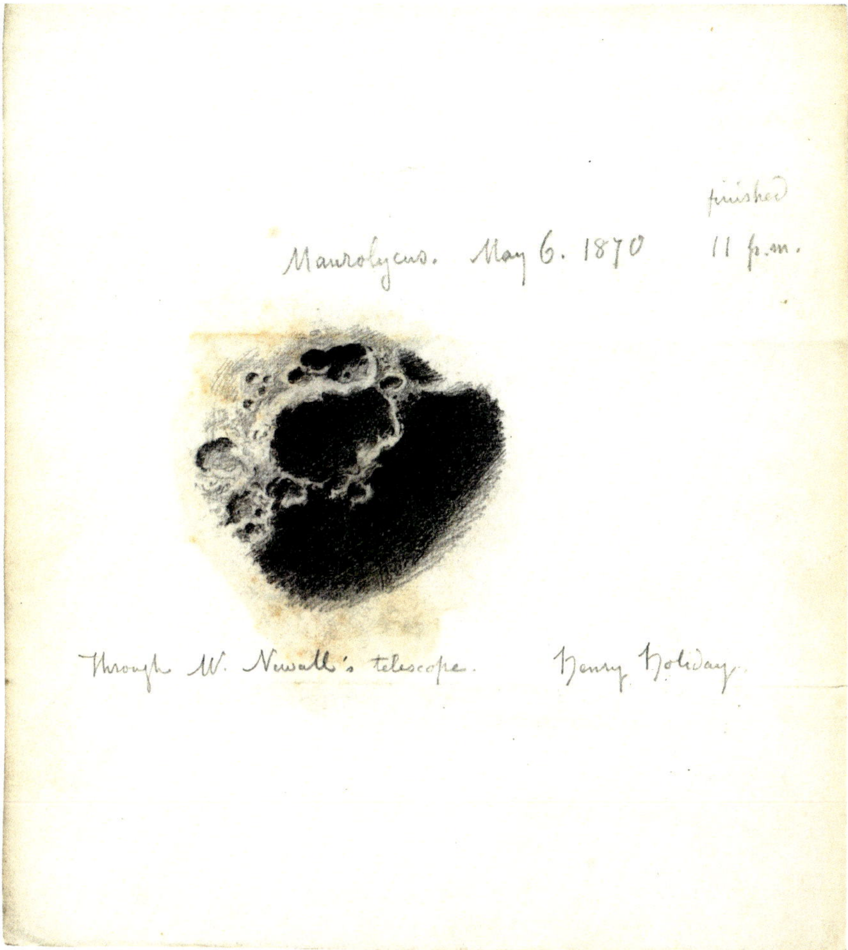

FIGURE 4.3 Maurolycus crater on the Moon, sketched by Henry Holiday using the Newall telescope, 1870
COURTESY OF CAMBRIDGE UNIVERSITY, INSTITUTE OF ASTRONOMY LIBRARY

Holiday visited the observatory again in April 1871 for two nights. On the first night, he completed a sketch of the Eratosthenes crater within four hours, this time using an eyepiece with 300× magnification (*Fig. 4.4*). Newall again manually moved the telescope, although this time it was because the clock drive needed some repair.[38]

38 Holiday, *op. cit.* (n. 22), p. 172. A few weeks later Holiday accepted an invitation from his friend, astronomer Norman Lockyer to accompany him to Ceylon (now Sri Lanka) to sketch the solar corona during the total eclipse of 12 December 1871.

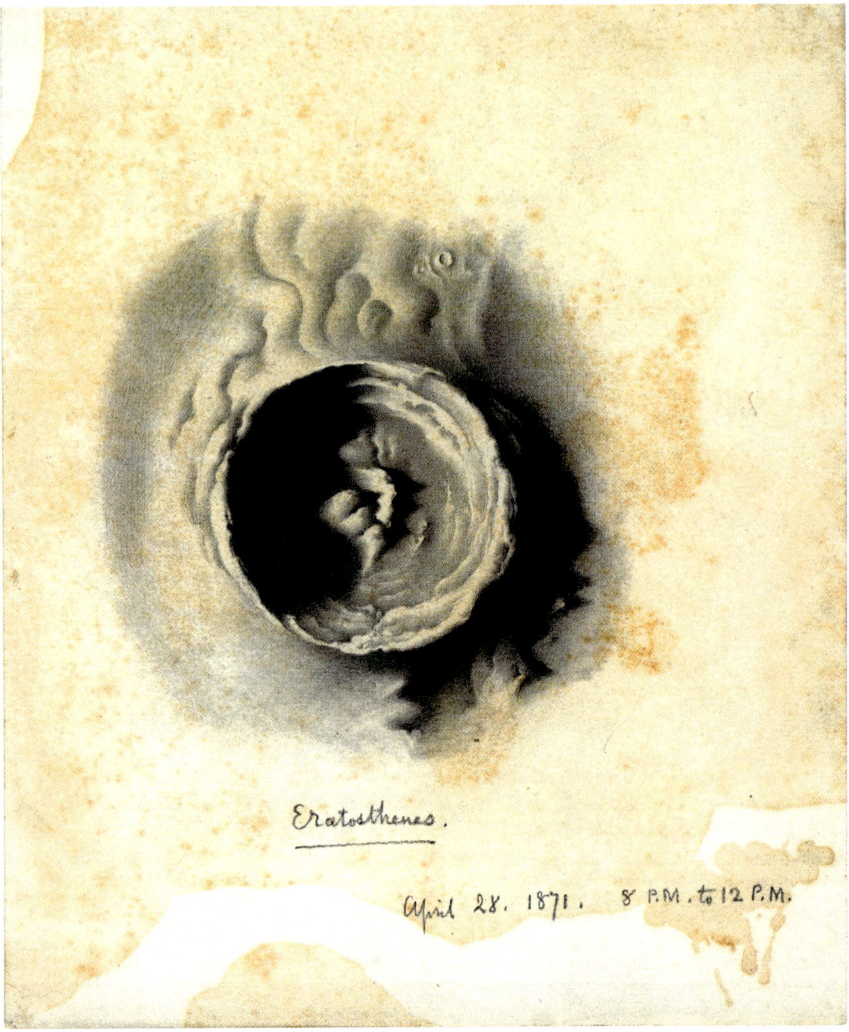

FIGURE 4.4 Eratosthenes crater on the Moon, sketched by Henry Holiday using the Newall telescope, 1871
COURTESY OF CAMBRIDGE UNIVERSITY, INSTITUTE OF ASTRONOMY LIBRARY

In spite of all the difficulties, Newall did use the telescope. He made systematic observations of a comet at the end of June 1881. Three of his drawings and a description of the comet were published.[39] In 1886 he commissioned a

39 W. H. M. Christie, "The Comet", *Nature*, 30 June 1881, pp. 197–200: 198–200.

"clever local optician" to build an occulter for the telescope, so that he could more easily observe faint satellites by shutting off the near planet.[40]

4 Second Stop: Cambridge

On 4 March 1889 Newall, already seriously ill, wrote to the Vice-Chancellor of the University of Cambridge to offer the telescope and its dome. A committee consisting of W. H. Christie (1845–1922), A. Ainslie Common (1841–1903), amateur astronomer and expert on astrophotography, and Andrew Graham (1815–1908), discoverer of the asteroid 9 Metis, visited the observatory on 28 and 29 March 1889. After examining the telescope and dome, the committee recommended the offer to the syndicate appointed by the university; it consisted of J. C. Adams (1819–1892), E. J. Routh (1831–1907), J. W. L. Glaisher (1848–1928), G. D. Liveing (1827–1924), G. H. Darwin (1845–1912), and J. J. Thomson (1856–1940).[41] The endorsement implies that the telescope remained of high quality even twenty years after its construction.

Newall died in 1889 while the university was deliberating. The university was wary because there were neither financial resources for the transport and erection of the telescope nor experienced personnel to use it. Newall's younger son solved these problems. Hugh Frank Newall (1857–1944) had been working as a junior demonstrator of physics in the Cavendish Laboratory of Cambridge University since 1885.[42] Wanting to make full use of the telescope, he offered to allocate the required amount of money and work for five years without salary as the main observer in charge of the telescope. His condition was that the university would provide him with space near the telescope to build a house.[43] The university accepted his terms, and the telescope was installed in Cambridge in late 1891.

---

40 R. S. Newall, "Newall's occulter", *Notices of the Royal Astronomical Society*, 47 (1886), pp. 33–34. doi: 10.1093/mnras/47.1.33.

41 W. H. Christie, A. Ainslie Common, and A. Graham, "Preliminary Report of the Newall Telescope Syndicate", *Nature*, 30 May 1889, pp. 114–115.

42 *A History of the Cavendish Laboratory 1871–1910*, Longmans, Green, and Co., London, New York, 1910, p. 331. H. F. Newall studied in Trinity College, Cambridge and then he worked as an assistant master in Wellington College. He spent some years in Europe until Professor J. J. Thomson invited him to work as his assistant in Cavendish Laboratory. Thomson and Newall published several papers together; see: E. A. Milne, "Obituary, Prof. H. F. Newall F.R.S.", *Nature*, no. 3885, 15 April 1944, pp. 455–457; also: Milne, *op. cit.* (n. 15), p. 720; the latter includes a full bibliography of Newall's work.

43 Milne, *op. cit.* (n. 15), p. 721.

In the following year, H. F. Newall prepared it for spectroscopic observations. He had already resigned his position in the laboratory in 1890[44] as he "consented to become responsible for the performance and superintendence of the researches which the acquisition of this fine instrument rendered possible".[45] With the installation in early 1893 of a new clock drive manufactured by the Grubb Telescope Company run by Sir Howard Grubb (1844–1931), and the acquisition of two objective prisms (10.5 inches in diameter and 7.5 degrees in angle) by the well-known astronomer Sir Norman Lockyer (1836–1920), Newall took photographs of Vega's spectrum. By 1895 it had become clear that the performance of the instrument with objective prisms was limited. A large dispersion spectrometer was needed. A new spectrometer was installed; it had one prism and gave a satisfactory dispersion.

Thus, from 1896, H. F. Newall, having overcome significant technical difficulties, began systematic spectroscopic observations to determine the radial velocities of various bright stars. The first results were satisfactory. In 1897 H. F. Newall proposed to Cambridge University to serve for another five years, without salary, as chief observer of the telescope.[46]

In 1899 the one-prism spectrograph was replaced by a new instrument consisting of four prisms. At the same time, H. F. Newall's observations led to significant results. He discovered, along with W. W. Campbell of the Lick Observatory, that the star Capella is a spectroscopic binary. He determined with admirable accuracy the mass and brilliance of the star. He also discovered that the star α Persi has a variable radial velocity.

In 1905, continuing his studies in stellar spectroscopy, he published a detailed description of the four-prism spectrometer, a new work on the radial velocities of stars, and an article, still considered a classic, on the design and construction of spectrographs. In this article H. F. Newall recounts all the experience he gained in trying to make good quality spectroscopic observations for twelve years with the telescope his father built. At the same time, he designed and built a new spectrograph without prisms. This new spectrograph used a diffraction grating and was put to work in late 1905. He also used the lens of the Newall telescope, after removing it from the tube, in combination with a two-mirror heliostat for spectroscopic observations of the Sun. After two years of observations, he published the first results of his observations of the Sun.

44 N. R. Campbell, "1903–1909", in *History of the Cavendish Laboratory*, *op. cit.* (n. 42), p. 223.

45 L. R. Wilberforce, "The Development of the Teaching of Physics," in *History of the Cavendish Laboratory*, *op. cit.* (n. 42), p. 256.

46 Milne, *op. cit.* (n. 15), p. 723.

In 1909 he was elected Professor of Astrophysics at the University of Cambridge, while at the same time he was director of the Solar Observatory of the same university. Because of his new duties and ever-increasing interest in solar physics, he spent most of his time neglecting his activities in night astronomy, especially after 1911. However, he observed the spectrum of Comet Halley in 1910, while his various assistants continued to photograph stellar spectra and collect valuable observations.

Newall retired in 1928 but continued to work scientifically until his death in 1944. H. F. Newall's research with the Newall telescope was innovative and resulted in several papers and an influential book on spectroscopy.[47] This is why he is considered to be among the most significant scientists in solar and stellar spectroscopy of his era.[48]

F. J. M. Stratton (1881–1960) was his successor in both the position of director of the Solar Observatory and professor of astrophysics at Cambridge University. He had been working with the Newall telescope since 1911, studying stellar radial velocities while observing the spectra of bright diffuse nebulae as well as Nova Lacertae 2010. In 1913 and 1914 Stratton published two papers on the spectral analysis of this nova. He continued various areas of research at the Cambridge Observatory, but old-fashioned instruments, such as the Newall telescope, were slowly abandoned.[49]

By the end of the 1950s, the Newall telescope was completely obsolete, and its future was dull. Its dome had also become really difficult to use. The time was ripe for the Newall telescope to travel to its next destination.

5 Third Stop: Athens

In 1955, George Kontopoulos (1928–), a Greek astronomer and later academician, participated in the General Assembly of the European Astronomical Union in Dublin. There he was informed that Cambridge University was willing to donate the Newall telescope to Greece if it were not wanted by a British

47 H. F. Newall, *The Spectroscope and Its Work*, Society for Promoting Christian Knowledge, London, 1910.

48 John Hearnshaw, "Newall, Hugh Frank" in Thomas Hockey (ed.), *Biographical Encyclopedia of Astronomers*, Springer, New York, 2014, pp. 1561–1562.

49 Nevertheless, spectroscopic observations continued for several years until the Second World War. Some papers were even published, such as the one on the Novae Herculis spectrum in 1935: F. J. M. Stratton and E. G. Williams, "Nova Herculis", *Nature* 135, no. 3417 (1935), pp. 657–657, doi:10.1038/135657c0.

observatory.[50] In August after the meeting, a group of Greek and British scientists examined the condition of the telescope and agreed that the unused telescope remained in good condition and had good optical properties. Given the high quality, it was worth finding the money to transfer the Newall telescope to Greece. The costs were covered by donations of Greek ship owners in London, especially by funds secured by N. Efstathiadis, director of A. M. Nomikos Maritime Company.[51] The telescope was dismantled in about 1956 by Sir Howard Grubb Parsons and Co. Ltd.,[52] and arrived in Piraeus on 16 June 1957. The transport was carried out by the P. Kallimanopoulos Hellenic Maritime Company. It took fifteen days for the instrument to reach the Penteli observatory, arriving on 2 July 1957.[53] (*Fig. 4.5*)

A new life for the Newall telescope began in Penteli at the National Observatory of Athens. The Greek state financed the construction of a new dome, the diameter of which was 18 metres (*Fig. 4.6*). The floor could be elevated up to five metres.[54] The telescope was ready to search the skies in November 1958. It was tested and found able to observe stars up to a magnitude of 16.2 at an altitude of more than 60°.[55]

Between 1961 and 1964 the telescope's equipment was enriched with the purchase of sixteen eyepieces by Zeiss,[56] a micrometer by R. Danger, a precision polarimeter by A. Jobin – G. Ynon,[57] and a filar micrometer by Morin (successor of Secretan), Paris.[58] The clock drive of the Newall telescope was repaired and its mechanism, which operated with weights, was replaced with an electric motor by a laboratory technician, E. Sigalas in 1963.[59]

---

50 Nikos Matsopoulos [Νίκος Ματσόπουλος], *Astronomy in Modern Greece* [*Η αστρονομία στη σύγχρονη Ελλάδα*], [self-published], Athens, 2000, p. 174. *Hellenic Military Geographical Service Bulletin*, nos. 69–70 (Semester 1–2, 1958), pp. 27–31.

51 Nikos Matsopoulos, "The Travels of an Impressive Refractor: The 25″ Newall", *Journal of the Antique Telescope Society* 8 (1995), pp. 8–11.

52 Dewhirst, *op. cit.* (n. 7), p. 495. For the remainder of the dome in Cambridge Observatory, see: Christopher Evans and Richard Newman, "An 'Imperial Philosophical Machine': The Archaeology of the Cambridge Observatory and Early Modern Science", *Antiquity* 85 (2011), pp. 1369–1384.

53 Stavros M. Plakidis (Σταύρος Μ. Πλακίδης), "The National Observatory of Athens and Its Work (1842–1965) [Το Εθνικόν Αστεροσκοπείον Αθηνών και το έργον του (1842–1965)]", *Technical Chronicles* [*Τεχνικά Χρονικά*] 8, no. 518 (1969), pp. 475–485: 482.

54 *Hellenic Military Geographical Service Bulletin*, no. 77 (Semester 1, 1960), p. 57.

55 *Hellenic Military Geographical Service Bulletin*, no. 81 (Semester 1, 1962), p. 88.

56 *Hellenic Military Geographical Service Bulletin*, no. 84 (Semester 2, 1963), pp. 201–220.

57 Plakidis, *op. cit.* (n. 53), p. 483.

58 Plakidis, *op. cit.* (n. 53), p. 483. Also, see: *Hellenic Military Geographical Service Bulletin*, nos. 85–86 (Semesters 1–2, 1964), p. 54. In this source Secretan is mentioned incorrectly as the maker of the instrument.

59 *Hellenic Military Geographical Service Bulletin*, no. 89 (Semester 1, 1966), p. 64.

FIGURE 4.5 The dismantling of the dome for the Newall telescope at Cambridge in 1957
COURTESY OF CAMBRIDGE UNIVERSITY, INSTITUTE OF ASTRONOMY LIBRARY

FIGURE 4.6 The dome of the Newall telescope at Penteli Observatory
COURTESY OF NIKOS MATSOPOULOS

The telescope was used to observe variable stars, the Sun, artificial satellites, and comets[60] as well as the solar eclipse of 20 May 1966.[61] It was also used

60 The Newall telescope was used to observe the comets Alcock (1959e) and Giacobini-Zinner (1959b): *Hellenic Military Geographical Service Bulletin* 89 (Semester 1, 1966), p. 64; Seki (1961f): *Hellenic Military Geographical Service Bulletin*, no. 84 (Semester 2, 1963), pp. 201–205; 41P/Tuttle-Giacobini-Kresák and Humason (1961e): *Hellenic Military Geographical Service Bulletin*, no. 85–86 (Semesters 1–2, 1964), p. 56.

61 *Hellenic Military Geographical Service Bulletin*, no. 93 (Semester 1, 1968), p. 41.

for planetary observations in cooperation with IAU Commission 16 (Physical Study of Planets and Satellites). As a result, more than 11,000 pictures of Venus, Mars, Jupiter, and Saturn were taken during the years 1964–1969,[62] and the diameters of planets and satellites were measured.[63]

Due to increased light pollution and the construction of the 1.2-metre diameter mirror telescope in Kryoneri, Corinth in 1975, the Newall telescope ceased to be used for cutting-edge observations. It continued to be used for the training of students of the Physics Department of the National and Kapodistrian University of Athens. It is not an exaggeration to say that the Newall telescope served as the entry point for many Greek astronomers into the world of modern observational astrophysics for almost four decades.

6 New Role, New Life

The Penteli Visitor Centre was inaugurated in 1995. The main concern of the Centre has been the dissemination of astronomical knowledge and discoveries in space physics to the general public, and especially to young people. This is achieved through seminars, lectures, and night observations with the Newall telescope. In addition, the Centre produces films with astronomical content, which are presented to visitors during the tours. These activities owe a lot to the personal attention of the astronomer Nikolaos Matsopoulos.

From September 1995 until today, more than 200,000 visitors from Greece and abroad have been guided to the Centre. The special educational programmes organised for morning and evening tours for schools are attended every year by students from about 250 schools of every educational level throughout Greece. The telescope has not only starred in astronomy but also in cinema. In 2003 it appeared at the beginning of the movie *A Touch of Spice*. After a major repair to both the telescope and the dome in 2013, a new life has opened for the legendary telescope (*Fig. 4.7*).

62 *Hellenic Military Geographical Service Bulletin*, 90 (Semester 2, 1966), pp. 82, 85; *Hellenic Military Geographical Service Bulletin*, no. 93 (Semester 1, 1968), p. 43; *Hellenic Military Geographical Service Bulletin*, no. 95 (Semester 1, 1969), pp. 52, 55; and *Hellenic Military Geographical Service Bulletin*, no. 97 (Semester 1, 1970), p. 44.

63 For a detailed bibliography see: Matsopoulos, *op. cit.* (n. 51), pp. 177–179.

FIGURE 4.7 The Newall telescope today
COURTESY OF NIKOS MATSOPOULOS

7 Conclusions

One may wonder what a failure is. Something that is not working? Something that has not been done? If we choose these definitions, the Newall telescope was not a failure. It was well made, and it worked properly in technical terms. But it never fulfilled the great expectations of its original owner. It was a kind of a white elephant. Unique and rare, the great telescope lived in an unfriendly environment. Moreover, the world and science moved faster than it could adapt. We must also consider that the failure of this impressive instrument lay in the failure of its owner to recognise that times had changed. Being a man of the upper classes and very conservative, Newall believed that he could control not only the life of his workers but also the fate of this telescope. The construction of the telescope was not a scientific necessity but the fulfilment of a personal vanity. And this fact was reflected in the way it was used.

Although winners are often stars, losers are sometimes even brighter than them. The Newall telescope is an affirmation of this rule. It was (and still is) a highly competent instrument, which was unlucky to spend too many years in the wrong place.

CHAPTER 5

Not All Bad: Rehabilitating Phrenology in Museums

Tacye Phillipson

History is, epigrammatically, written by the winners. Science is too. At the heart of the scientific method is the discarding of hypotheses and ideas which are disproven, leaving the winners, until they are themselves perhaps disproven or superseded. The disadvantages of a neglect of the stories and viewpoints of the "losers" in history are manifest and acknowledged, and considerable effort is now devoted to reducing the imbalance. Less focus has so far been devoted to the way in which the many disproved or superseded hypotheses of science are neglected or negatively portrayed, especially when the history of science is talked about by scientists, and the implications of this for the current field of science.

It has been widely pointed out that science publication has an increasing problem with the disproportionate publication of positive results, leaving research which has negative results without credit and at risk of wasting resources by being repeated. This affects the way science is practiced and it is possible that a more positive portrayal of historic disproven hypotheses could be part of a valuable culture change in current science.

Phrenology – determining someone's mental propensities and abilities from the shape of their skull – is nowadays comprehensively dismissed as a failure, and more than that, as pseudoscientific nonsense. In this essay I shall use it as an example to argue for the importance of failures having a prominent place in the history of science and discussions of what science is. Further, I shall briefly highlight a couple of examples where current research addresses similar questions to historical phrenology, but phrenology's notoriety as a complete failure may obscure the nuances of its potential relationship to modern science.

Phrenology was developed by the German anatomist Franz Joseph Gall (1758–1828) and had its origins in some very fine dissection of the brain which distinguished between grey and white matter and led to his conception that different regions of the brain served different purposes. This idea is indeed accepted as correct, though Gall's assignment of specific purposes to particular regions was found to be spurious. Gall and his student Johann Kaspar Spurzheim (1776–1832), who coined the term phrenology, popularised their theory that as the cranial cavity is filled by the brain, the shape of the brain and size of its regions can be studied by examining the exterior of the skull,

 | DOI:10.1163/9789004689107_007

thus getting an indication of the shape of the cranial cavity in living people before the advent of x-rays or other medical imaging techniques. The shape of the skull, which was either felt with the fingers or measured "scientifically" with callipers, allowed a phrenologist to determine which areas of the brain were largest and most developed and thus argue that the purposes they served were most prominent in a person's character – size being equated to power. This found many advocates over the world, notably including George Combe (1788–1858) in Edinburgh, and the Edinburgh Phrenological Society, which he founded in 1820, had its rooms just across Chambers Street from where the National Museum of Scotland is located nowadays. It also attracted widespread popular interest, and Combe's phrenological book, *The Constitution of Man*, first published in 1828, was one of the best-selling books of the nineteenth century. Phrenological reasoning was also used in justification of racial supremacy arguments which were common at that time. The popular interest and general debate about phrenology survived long after it had ceased to have much serious consideration in mainstream science. (*Figs. 5.1, 5.2*)

During development of the National Museums Scotland's *Enquire* gallery, which opened in 2016, looking at science with an emphasis on the process of scientific research and advancement, I needed to justify on several occasions my proposed inclusion of phrenological material. This features as part of a display on medical imaging and research into understanding the brain. It is paired with a model of the white matter, or neural pathways in a brain, laser etched into glass – our way of portraying the Lothian Birth Cohort project, one ongoing current research project into cognition and the aging brain.[1] The phrenological head, showing which areas of the brain were associated with which mental abilities and the callipers for measuring the outside of the skull were part of a research question very similar to the modern project – why are people different? What can I find out about myself? But research in the early nineteenth century was limited by the tools available at the time. I argued that not despite, but because of, phrenology's lack of success in the study of the brain, it deserved a place in the display.

Phrenology was a failure, and it is important to include failures in the history of science. A history of science that only talks about the successes and accepted theories is a poor background for the process of doing science. The scientific method, as it is often simply described, includes formulating a hypothesis that fits existing data and theories. This hypothesis is rigorously

1 Ian J. Deary *et al.*, "The Lothian Birth Cohort 1936: A Study to Examine Influences on Cognitive Ageing from Age 11 to Age 70 and Beyond", BMC *Geriatrics* 7, no. 28 (2007), doi:10.1186/1471-2318-7-28.

FIGURE 5.1 Cast plaster phrenological head with numbered sections cast into the head, made by Francisco Mitchelelle of Pathhead, Scotland, mid-nineteenth century
NMS.T.1984.146. IMAGE © NATIONAL MUSEUMS SCOTLAND

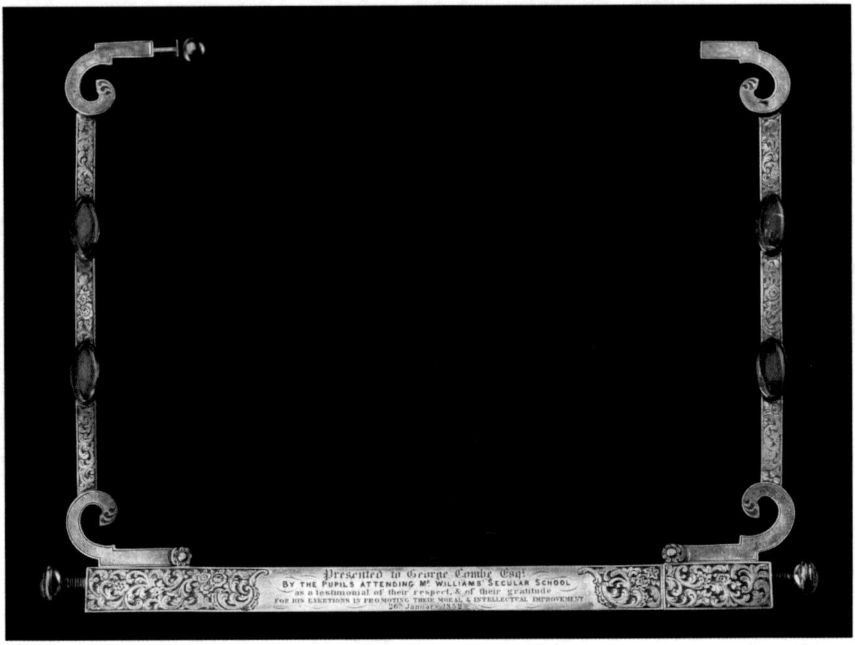

FIGURE 5.2 Set of silver phrenological callipers, presented to George Combe by the pupils of Mr Williams' Secular School, made by Michael Crichton, Edinburgh, 1852
NMS.T.2006.156. IMAGE © NATIONAL MUSEUMS SCOTLAND

tested and either supported, altered, or discarded accordingly. The progress of science is crucially dependent on these failures and the likelihood is that anyone undertaking a career in science will experience both success and failure – and considerable quantities of the latter. This is perhaps exemplified by Gall himself and his pioneering success at neuroanatomy and failure with phrenology. Portraying failure as a key part of the activity of doing science is necessary for giving a balanced view of the field, especially to young people who are not yet fixed in their career choices but may not feel they will be the one to formulate a great success, like the equations of electromagnetism, or predict a new subatomic particle. Also key to the scientific method is responding to data and being willing to change your mind in response to data. Gall was not at all exemplary here, preferring to explain away discrepancies. He is not the only scientist whose science suffered from this human weakness.

Phrenology, with its enduring popular appeal and discussion (like astrology today, I wonder if it was perhaps widespread as entertainment and not always sincerely believed in) is also a characteristic representative of pseudoscience. It has a basis in a mix of some good and robust science with other incorrect statements combined with a wish for there to be some solution to the problem that it purports to address. Awareness of what pseudoscience is, and of scientific-sounding language being used about theories that are not scientifically accepted, is of significant current relevance and importance. That phrenology is seen as historical and no longer has any significant following makes it a less fraught topic to use to introduce discussion and thinking about what pseudoscience is, and how to identify it, than any of the many current examples.

In the case of the *Enquire* gallery we decided that phrenology deserved its place in the narrative, as part of our efforts to portray the process of doing science, with all its complication, dead ends, wishful thinking, and involved personalities.

Phrenology was mainly but not exclusively applied to humans. There were some attempts to invoke phrenology to explain the characteristics of animals (*Fig. 5.3*). This leads to the first of my examples of modern science using ideas that I think the old phrenologists would recognise. While in studies of human cognition there are many techniques available, notably including an increasing range of functional brain imaging techniques, these are obviously not possible when studying extinct animals. Their brains no longer survive, but in some cases a well fossilised skull permits modern scanning to reconstruct the external shape of the brain. Studies comparing the gross morphology of brains of modern animals, such as the relative size of olfactory lobes or visual cortex, to their abilities can support deductions made from the skulls and hence brain shapes of extinct animals.[2] Correlating the size of brain regions with mental abilities is not now called phrenology, and this research will hopefully avoid the pitfalls into which its predecessor of the nineteenth century fell.

In humans, another topic of modern research and concern that is related to skull shape and mental abilities is plagiocephaly, or flat headedness, in babies. This has become of increasingly common concern since 1989 when it became known that babies sleeping on their fronts were more likely to experience cot death, which led caregivers to increase the average time babies spent lying on their backs. There is increasing evidence of a correlation between plagiocephaly and broader developmental differences, and research is ongoing into verifying these observations and elucidating the nature of any possible causation. Is the plagiocephaly an indication of a less mobile baby spending more time in one position? Does the plagiocephaly hinder development purely mechanically by increasing resistance to head movement? Or does the skull shape itself affect the brain's development? A search on the PubMed database

2 Stigg A. Walsh and Monja A. Knoll, "Directions in Palaeoneurology", *Special Papers in Palaeontology* 86 (2011), pp. 263–279.

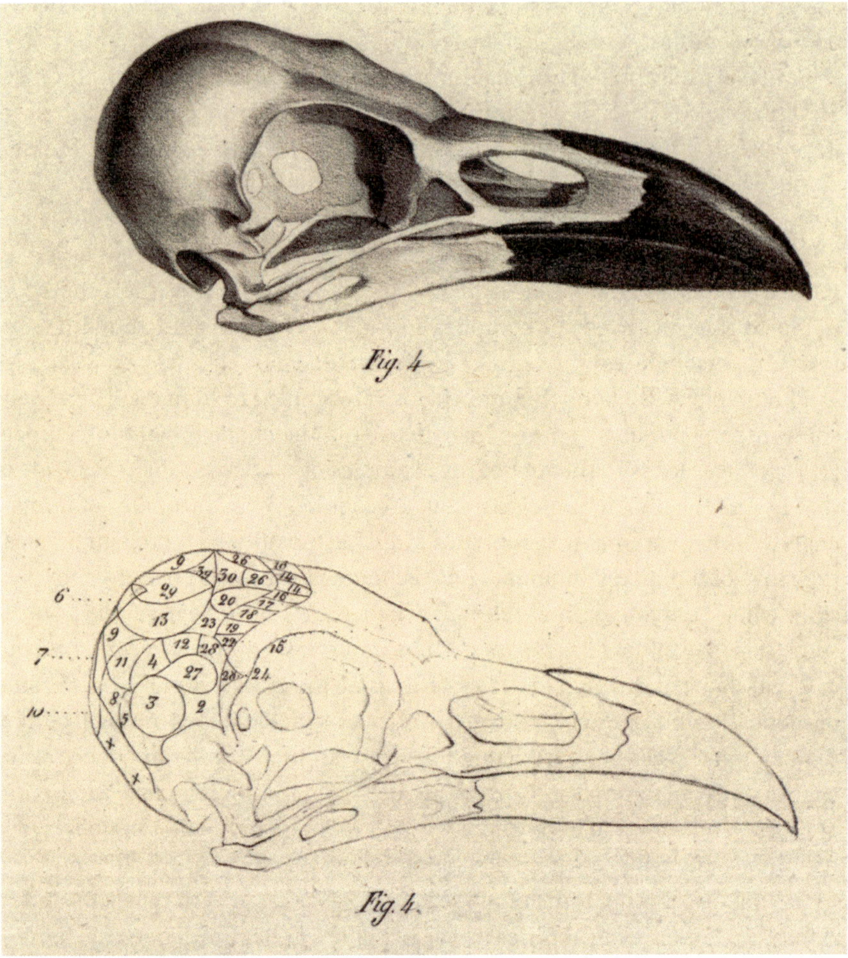

FIGURE 5.3 Phrenology of the crow, from Joseph Vimont's *Traité de phrénologie humaine et comparée*, 1831
COURTESY OF THE UNITED STATES NATIONAL LIBRARY OF MEDICINE

for the terms "plagiocephaly" and "cognitive" results in thirty papers published since the year 2000.[3] None of these papers also include the key word phrenology, unsurprisingly.

The completeness with which phrenology has been judged a failure by the light of science and history is perhaps a reflection of its historic success. As a topic of widespread interest, it attracted considerable attention and denunciation. It also enhances the perceived distance between the failure of the past and

3 PubMed, https://pubmed.ncbi.nlm.nih.gov/.

present scientific research despite similarities in the questions asked, barriers to answering them, and even in approaches taken. But, like so many stories of the progress of science it has much to reveal, both positive and negative, showing the field of science as more nuanced, less idealised and resultantly more interesting than a pure triumphalist development and accretion of accepted scientific truths with the digressions ruthlessly pruned from the narrative. An awareness of the importance of failure and the lure of pseudoscience are necessary for general scientific literacy and an integral part of the story of science. As such, phrenology justifies its place in our science gallery alongside great successes and ongoing research which will play its part in scientific progress, whether it is ultimately successful or not.

CHAPTER 6

Finding Way at Sea – Alternatives to the Log-Line and Sand Glass. The British Experience in the Hanoverian Era

Conservative Navigational Practice or Technological Failure?

D.J. Bryden

1 Navigational Background – Heaving the Log

The log and line for measuring speed and hence distance run, appears to have been an invention of English mariners, described in print by William Bourne in 1574, beginning to come into more general use from the 1590s.[1] Use of the log and line underpinned navigation by "dead reckoning", and despite significant advances in scientific navigation through increasingly precise celestial observation, the technique continued beyond the age of sail and well into the Victorian era. Heaving the log was a relatively simple and routine ship-board activity. The log was thrown overboard on the lee-side towards the stern, and the log-line, wound on a free running reel, let run until the log was clear of the effect of the ship's wake. The end of the initial length of "stray line" was usually marked with red rag at 10 fathoms (60 feet or 18.3 metres but might be longer in larger vessels); as that mark passed the side of the vessel the officer shouted "turn", and the half-minute sand glass was set going; when it ran out the officer called "stop". The line was then stopped. The length of line from the rag mark indicated the distance run in that half minute. To assist computation, the line was marked in intervals with "a piece of Twine with Knots in it, inreeved between the Strands of the *Line*, which shews how many of these spaces, or *Knots*, are run out during the half Minute".[2] The log was "generally about a quarter of an inch thick, and five or six inches [6 inches = 152.4 mm] from the angular point to the circumference. It is balanced by a thin plate of lead, nailed upon the arch, so as to swim perpendicularly to the water, with

1 David W. Waters, *The Art of Navigation in England in Elizabethan and Early Stuart Times*, Holllis and Carter, London, 1958, pp. 122, 137–138, 425–435. Eva G. R. Taylor (ed.), *A Regiment for the Sea and Other Writings on Navigation by William Bourne*, Cambridge University Press for the Hakluyt Society, 1963, pp. 126–128, 237, 297–299.

2 Archibald Patoun, *A Compleat Treatise of Practical Navigation*, J. Brotherton *et al.*, London, 1734, p. 143.

 | DOI:10.1163/9789004689107_008

about $^2/_3$ impressed under the surface."[3] The log-line was attached to the log by a three point harness, designed such that one strand of the harness pulled out a pin set in the log when the line was stopped. This made the log swim on its face, allowing it to be dragged back on board with reduced resistance.

Textbooks throughout the eighteenth and well into the nineteenth century include discussion of "heaving the log" as central to navigation by "dead reckoning".[4] All of them point to long known weaknesses of the method. One basic flaw was the division of the log-line, with different suppositions made over the period as to the relationship of the mile to a degree. For example, the earlier editions of John Hamilton Moore's long running textbook record:

> Mr. Norwood, and several other able Mathematicians, have found that a Degree of a great Circle upon the Earth contains about 367200 English Feet, therefore a Nautical Mile being the $^1/_{60}$ Part of 367200 Feet, that is, 6120 Feet, and since Half a Minute is $^1/_{120}$ Part of an Hour, the Length of a Knot on the Log-Line ought to be the $^1/_{120}$ Part of 6120 Feet, or 51 Feet. But as for the most Part the Ship's Way is found, by Experience, to be really more than that given by the Log, and as it is safer to have the reckoning before the Ship than after it, therefore 50 Feet may be taken as the proper Length of each Knot, and these Knots subdivided into ten Fathoms, each of 5 Feet, which is certainly the best adapted for practice, and will correspond with all the Tables and Instruments used in Navigation, as they are decimally divided, and, consequently, the Ship's Run determined with greater Ease and Certainty.[5]

3 Art. "Log" in William Falconer, *An Universal Dictionary of the Marine*, new corrected edn., T. Cadell, London, 1780, sig. 2A3^v–2A4^r.

4 For example: John Seller, *Practical Navigation*, 4th edn., J. Darby, London, 1680, pp. 224–226 and subsequent editions. Art. "log-line" in *The Gentleman's Dictionary … from the sixteenth edition of the original French, … with large additions, alterations and improvements adapted to the customs and actions of the English*, H. Bonwicke *et al.*, London, 1705, sig.3D8^r–^{v}. Patoun, *op. cit.* (n. 2), pp. 143–150 ran to eight editions by 1770, and was used by the Academy in Tower Street, London. John Hamilton Moore, *The Practical Navigator*, Richardson & Urquhart, London, 1772; as *New Practical Navigator* from 10th enlarged edn., B. Law & Son, London,1794; from 18th edn., enlarged and improved by Joseph Dessiou, F. & C. Rivington *et al.*, London, 1810; to the final 20th enlarged and corrected edn, (ed.) Joseph Foss Dessiou, T. Cadell *et al.*, London, 1828; the treatment of the log-line and dead reckoning remained substantially unchanged. The same is the case for John William Norie, *A new and complete Epitome of Practical Navigation*, for the author and William Heather, London 1805, pp. 62–63; the entry remains at least up to the 20th edition of 1872!

5 Moore, *op. cit.* (n. 4), 3rd edn., B. Law *et al.*, London 1784, pp. 177–178; 8th edn., B. Law *et al.*, London, 1807, pp. 177–178.

Later editions edited by the experienced navigator and hydrographer Joseph Dessiou, add:

> But some experienced Commanders find, that the allowing 50 Feet to a Knot, generally makes the Ship a-head of the Reckoning; and, to avoid Danger, mostly divide the Log-Line into Knots of 7 or 7½ Fathoms, of 6 Feet each, to correspond with a Glass that runs 28 Seconds. Others, again, divide the Seconds the Glass runs by 4, and take the Quotient for the Distance in Fathoms between the Knots; which of these Methods are best; I leave to every Captain's own Experience to determine; but certain it is, that whatever Length the Knots are, the most convenient Way is to divide them into Tenths.[6]

In the technical dictionary compiled in the first decade of the eighteenth century by John Harris FRS, the entry on the log-line closes by contrasting Norwood's computation of the size of a degree and thus the length of a log-line for use with a 30 second glass, with the error introduced by navigators who use a log-line with 42 feet or seven fathoms between each knot. Such users "correct" by using a 24 or 25 second glass. Of this practice Harris commented:

> But this is nothing but correcting one Blunder or Error by another; and shews plainly that the Common Sailors will not go out of their way, tho' they are sure they are in the wrong.[7]

The accuracy of the half-minute glass was another problematic area. On the one hand, it was widely recognised that the flow rate was impacted by damp weather.[8] The standard suggestion was that the glass should be checked against either a watch, or by setting up a seconds pendulum, though whether this was to be 39 $^1/_6$, 39 $^1/_5$, 39 ¼, or 39 ½ inches long varied between authorities.[9] Such a seconds pendulum might provide a tolerable check at sea in a flat calm. The knowing navigator took a battery of time-glasses on board, and would routinely

6 Moore/Dessiou, *op. cit.* (n. 4),18th edn., p. 132.

7 Art. "Log-line" in John Harris, *Lexicon Technicum, Or, An Universal English Dictionary of Arts and Sciences*, Dan. Brown, *et al.*, London, 1704, sig .5B4^v–5C1^r.

8 Patoun, *op. cit.* (n. 2), p. 149: "*Half-Minute* or *Quarter-Minute-Glasses* ... are seldom or never true, because dry and wet Weather have a great influence on them; so that at one Time they may run more, and at another Time fewer than 30 Seconds, and t'is evident that a small Error in the *Glass*, will cause a sensible one in the Ship's Way."

9 $^1/_6$ Patoun, *op. cit.* (n. 2), p. 127; $^1/_5$ Robert Heath, *Astronomia Accurata; or the Royal Astronomer and Navigator*, for the author, London, 1760, p. 309; ¼ Moore, *op. cit.* (n. 4), 3rd edn., 1776, p. 154 and Norie, *op. cit.* (n. 4); ½ William Falconer, *An Universal Dictionary of the Marine*, new edn., ed. William Burney, T. Cadell & W. Davies, London, 1815, p. 237.

check them against each other. Nevertheless, much depended on the manufacturer. An experienced naval ship's Master, like William Nichelson, noted:

> People too frequently take it for granted, that the half and quarter minute glasses are the just length they are marked upon the end, or sometimes both ends, with a pen and ink, 28, 29 or 30 seconds, which I have reason to believe are marked by guess, without any regard to the true time, for which reason I have always made it a rule to try and prove the glasses before the ship sailed out of port, to proceed on her voyage to sea, and scarce ever found a glass marked right. I have seen one marked 27 or 28 seconds, which I have found upon trial to be full 30, and others that have been marked 30 seconds have not run more than 28.[10]

The log-line when new was likely to shrink considerably after being immersed – and even when in service was subject to stretching and shrinkage. Archibald Patoun, who became an Ordnance Officer, and may have had little direct experience of the realities of on-board navigation, noted:

> to be the better assur'd of the Ship's Way by the *Log-Line*, we ought to measure and alter the *Knots* on it every time before we use it; but this is seldom done oftener than once a Week, and sometimes not above once or twice in a whole voyage; also when the Line is measured so its greatest Degree of shrinking, it is generally left there; and when by much use, it comes to stretch again it is seldom or never mended, tho' it will stretch beyond what it first shrunk.[11]

---

10 William Nichelson, *A Treatise on practical Navigation and Seamanship*, for the author, London, 1792; cited from the 1796 printing when reviewed in *The Critical Review or Annals of Literature*, 24 (1798), pp. 142–143. Nichelson, serving as Master RN, had earlier published *Sundry remarks and observations made in a voyage to the East-Indies, on board His Majesty's Ship the Elizabeth, From the Beginning of the Year 1758, to 1764*, 2nd edn., W. Herbert, London, 1773. He was appointed Master Attendant at Portsmouth in 1770: see The National Archives [TNA]: ADM 106/1191/239; and in post in 1782 when the *Royal George* attempted a manoeuvre (which Nichelson had warned the Admiral against) to effect a repair to a sea-cock; see Hilary L. Rubinstein, *Catastrophe at Spithead: the sinking of the Royal George*, Seaforth Publishing, Barnsley, 2020, pp. 85–87. He was Governor and Master Attendant at Chatham from 1783 to at least 1797: see Parliamentary Papers, *Reports from Committees of the House of Commons*, 13, 34th Finance Report 1798, "Appendix to Chatham Chest", London, 1803, p. 598.

11 Patoun, *op. cit.* (n. 2), p. 150. Moore, *op. cit.* (n. 4), 3rd edn., B. Law *et al* London, 1776, p. 178, is less specific "Care should also be taken to measure the Log-Line pretty often, less it stretch, and deceive you in the Distance." John Bettesworth, *The Seaman's sure guide, … for the use of schools*, for the author, London, 1783, p. 29 would have the line measured "very often". Andrew Mackay, *The Complete Navigator*, Longman & Rees, London, 1804, p. 7 in

Overall, it was agreed that the experienced navigator would routinely adjust the daily distance run from dead reckoning based on hourly heaving the log, to take account of fluctuations in the strength and direction of the wind, of currents, and especially when running before the wind, when the sea would tend to "bring home the log". In 1623, buccaneer and subsequent naval officer Sir Henry Mainwaring, in his *Sea-mans Dictionary*, says of the log and line: "this is a way of no certainty unless the wind and sea and the course continue all one, besides the error of turning the glass and stopping the line".[12] Patoun concluded that the whole method was "subject to several Errors, and these pretty considerable".[13] William Burney, who founded a successful naval academy at Gosport, amended the entry in Falconer's *Dictionary of the Marine*, adding the warning:

> The log, however, is a very precarious way of computing, and must always be corrected by experience and good sense; there being a great deal of uncertainty in the yawing of the ship going with the wind aft, or upon the quarter, in the heaving of it, by its coming home, or being drawn after the ship; on account of the friction of the reel and lightness of the log in the course of the current, and in the strength of the wind, which seldom keeps the same tenor for two hours together.[14]

To the scientific navigator, "dead reckoning" was a last resort; turned to only when it proved difficult or impossible to fix one's position using celestial observations. Yet the technique remained at the core of day-to-day operations. Above all it was a simple, routine measurement undertaken as a matter of course every hour on naval and East India Company ships, and recommended every two hours on merchant vessels. It required only basic numerate skills,

---

a footnote adds: "A log-line constructed of cotton yarn will not alter in length". Adam Glendenning, of the Naval Hospital at Yarmouth, who used Mackay as a prime source, noted: "cotton lines have been recommended, and particularly adopted, as being less liable to vary in length"; see John Mason Good, Olinthus Gregory and R. Newton Bosworth (eds.), *Pantologia: a new Cyclopaedia*, London, 1808–1813, 12 vols.: v. 8: art. "Navigation", sig. Z6[v] – explicitly attributed to Glendenning in *ibid.* 1 (1813), sig.b3[r]. I find no other evidence that Mackay's recommendation influenced custom and practice.

12 The published edition was published by John Bellamy, London 1644, but the manuscript was completed in 1623, with copies circulating widely in manuscript: see G. E. Mainwaring (ed.), *Life and Works of Sir Henry Mainwaring* Naval Records Society, London, 1922, 2 vols.: v. 2, pp. 69–82. Quoted from Waters, *op. cit.* (n. 1), pp. 137–138.

13 Patoun, *op. cit.* (n. 2), p. 149.

14 Falconer/Burney, *op. cit.* (n. 9), p. 238. On the academy see A. Macdermott, "Dr Burney's Royal Academy at Gosport", *Mariner's Mirror* 51 (1965), pp. 55–57.

though the day's run might well be adjusted by an experienced master. The mathematically competent navigator would rely on latitude and longitude computations using the increasingly sophisticated instruments, calculations, and tables – reverting to "dead reckoning" when the weather conditions precluded making the requisite solar, lunar, or stellar observations. Given this situation, it is not surprising that throughout the Hanoverian period, there was a succession of attempts to improve on "dead reckoning".

In 1773 the Hon. Constantine Phipps RN (1744–1792), active in parliament and appointed a Lord of the Admiralty in 1777, commanded HMS *Racehorse* on a polar expedition, proposed by the Royal Society. Phipps wrote a thorough account of the voyage, making perceptive observations on the instruments he had been instructed to trial. He also surveyed the literature dealing with methods of measuring a ship's way, expressing particular surprise over inconsistencies between various authorities and naval custom and practice in the spacing of knots on the log line. Phipps enumerated five reasons for inaccuracies in measuring the ship's way with the log and line:

1. current
2. yawing of the ship
3. swell
4. coming home (that is the log being drawn towards the ship by friction in the reel, or a following sea)
5. speed averaged from hourly readings, with the officer of the watch making an allowance for his estimation of variations in wind force over the period "for which an arbitrary correction is made by the officer of the watch; and though men of skill and experience come near the truth, yet this allowance must, from its nature, be inaccurate".

Phipps supervised a series of comparisons between the conventional log and line and the conical log with sinker made after the 1747 design of the French mathematician and "hydrographe royal" for Le Croisic and Le Havre, Pierre Bouguer. Phipps considered it to be "extraordinary that this log should never have been made use of by others"; but Phipps was a widely read Fellow of the Royal Society, had access to *Mémoires* of the French academy, and to the most recent edition of Bouguer's textbook on navigation. Only after the publication of his account of the polar expedition does Bouguer's design begin to be mentioned in some English language publications – though it gets no place in navigation textbooks. Phipps concluded that the French design was to be preferred, since the sinker ensured the conical log swam at a height that minimised the effect of wind. Nevertheless, his conclusion was cautiously moderated by the caveat of an experienced naval commander: "it is not to be determined that the observations of a single voyage can be sufficient to determine the merit of any instrument, particularly one of so much consequence as the log".[15]

2 Alternatives to the Log-Line and Half-Minute Glass

Given that even teachers of the practice of basic navigation were quite open about the failings of the log-line and half-minute glass to provide location by "dead reckoning", it is not surprising to find many alternative log designs being proposed in the Hanoverian era. A list of these, by date, is appended. Even the theoretical expression of an idea, written by a designer with little or no first-hand experience of the sea, underlines awareness that conventional custom and practice was unreliable. It is useful to consider the impact of these alternative designs at various levels, before considering why, with very few exceptions, they can be judged to have failed to change long standing on-board custom and practice.

3 Literary Impact

The compilers of dictionaries and encyclopaedias, like many writers of textbooks, heavily recycled existing printed material. New designs that were published in the periodical press were readily available to supplement entries that described dead reckoning under the entry "log-line" or "rate of sailing". For example, Andrew Wilson, an un-placed naval lieutenant, compiled and published a naval history, inserted a notice of Gould's perpetual log (1800) twice, each taken from a different source,[16] but Wilson had no first-hand knowledge

15 Constantine John Phipps, *A voyage towards the North Pole undertaken by his Majesty's command 1773*, J. Nourse, London, 1774, pp. 87–94. See also Ann Savours, "'A very interesting point in geography': the 1773 Phipps expedition towards the North Pole", *Arctic* 37 (1984), pp. 402–428. For Bouguer's conical log design see Pierre Bouguer, "Sur une nouvelle construction de Loch et autres instrumens qui peuvent servir à mesurer le sillage des vaisseaux", *Mémoires de l'Académie Royale des Sciences*, Paris 1747, pp. 644–664 is the source cited by Phipps. Pierre Bouguer, *Nouveau Traité de navigation*, H.-L. Guerin & L.-F.Delatour, Paris, 1753, pp. 98–110, adds a spring-loaded design which was intended to give an instantaneous on-board reading. Phipps was also aware of the account of the conical log and sinker in Bouguer, *op. cit.* nouvelle edition, revu et abrégé N. de la Caille, L. Delatour, Paris, 1769, pp. 73–82; but he did not comment on the later spring-loaded design, which I judge to be a largely theoretical proposal of a savant with expertise in hydrodynamics.

16 Andrew Wilson, *Appendix to the Naval History of the United Kingdom*, [?for the author], Cork, 1807, p. 127 reprints "Perpetual Sea-Log", from *Union Magazine and Imperial*

of the device. The Edinburgh-based *Encyclopaedia Britannica* inserted an account of Gottlieb's perpetual log (1791) into the third edition of 1797 – but the editors had also included accounts of that inventor's patents for cranes (1786) and axle-trees (1787) – almost certainly because Gottlieb had provided the information as part of his marketing activity.[17] The twelve-volume London published *Pantologia* (1813) supplemented a standard derivative entry on the log-line, with details of the alternative log designs of Bouguer, plus those of Russell (*c.*1750) and Foxon (1772), citing Phipps for all three, and Gottlieb (1791); then follows descriptions of the designs of Saumarez (1717) and Smeaton (1754) from earlier literature, plus Hamilton (1801), Hamill (1806), Leguin (1787), Boswell (1807), and Grant (1807), culled from contemporary sources. There is, however, no judgement on the efficacy or otherwise of any of these alternative designs, beyond the brief, and equally derivative comment on Grant's hydroscope, that the method "as well as Mr. Boswell's is open to many objections, which we have not room to specify here".[18] There is even wider trawling of contemporary literature in the somewhat later multi-volume *Encylopaedia Metropolitana* which appeared in parts from 1817. Two features distinguish this entry from the passive repetition of similar compilations. Firstly, designs using similar operational principles are grouped together. Secondly, there is an explicit understanding of the circumstances under which the log is used at sea.[19]

Of contemporary authors, Phipps, as noted above, drew attention to Bouguer's conical log and sinker, which he presumably had had made for him,

---

Register 2 (1801), p. 196; whilst later (pp. 149–150) he reprints "Description of a perpetual sea log, invented by Mr. Gould", from *Monthly Magazine or British Register* 15 (1803), p. 513; neither entry acknowledges the source.

17 D. J. Bryden, "Valentine Gottlieb, Immigrant Engineer of Lambeth: His Trade Card of c.1810 Unpacked", *Science Museum Group Journal*, issue 15 (2021); on line only, at http://journal.sciencemuseum.ac.uk/issues/spring-2021; Gottlieb's log was illustrated and described in "Explanation of a machine for ascertaining a ship's rate of sailing at sea, with a time regulator and tell-tale", *European Magazine* 20 (1791), p. 491; and republished in the influential *Collection of Papers on Naval Architecture, originally communicated through the channel of the European Magazine*, 1 part 2, J. Sewell, London, 1792, sig.D8^{r}–^{v} as an appendix to the reprinting of Joseph Gilmore, "The Improvement of Navigation by two new-invented engines: the one called a Navium, for measuring a ship's way, … first printed in 1722".

18 Art. "Log" in Good *et al* (n. 11), v. 7, sig.H7^{r}. The "objections" had been discussed in *Retrospect of Philosophical, Mechanical, Chemical and Agricultural Discoveries* 3 (1807), p. 360 on Boswell, and *ibid.* 5 (1810), pp. 57–60 on Grant.

19 Art. "Log" in Edward Sedley, Hugh J. Rose and Henry J.Rose (eds.) *Encyclopaedia Metropolitana; or a universal dictionary of knowledge*, R. Fenner, London, 1818–1845, 25 vols.; from 1821 J. Mawman *et al.*, and by 1845, B. Fellowes *et al*: v. 21 (1845), pp. 521–3; this volume "Miscellaneous and Lexicographical", v. 8, first appeared in 1832.

and at the specific request of the Admiralty, he trialled the perpetual logs of Russell and Foxon, which will be discussed later. The French Royalist refugee, Charles Grant, vicomte de Vaux, who first submitted his own perpetual log design to the Board of Longitude in 1806 (*Fig. 6.1*) and included it in his broad patent of 1807, put his design into context by describing, and largely belittling, such other logs as he knew of from his voracious reading. Citing Saverien's *Dictionnaire* he started with that of Vitruvius in classical antiquity, through the hydrodynamic designs of Poleni and Pitot. Drawing on Phipps he mentions the perpetual log designs of Saumarez (1717), Russell (*c.*1750) and Foxon (1772), and in a footnote that betrays the inadequacy of his technical comprehension, dismisses Massey's patent log (1802) as "an exact copy of what has been described in the above three instruments". Gilmore (1722) and Gottlieb (1791) get a mention; and having misread his sources, Grant introduced a canard into the literature with the "nautical dromometer" of Benjamin Martin.[20] Literary references need to be treated with caution, as the case of the French designer Estienne Leguin demonstrates. Leguin having failed to interest his countrymen in his log design, brought it to England, and submitted it to the Society of Arts in 1787

20 Charles Grant, *The means of finding the longitude at sea gradually developed, discovered*, for the author, London, 1808; see the separately paginated "Fourth Dissertation, on the mathematical, mechanical and astronomical means for finding the longitude at sea", pp. 7–19. John R. Millburn, *Benjamin Martin: Author, Instrument-Maker, and "Country Showman"*, Noordhoff International, Leyden, 1976, pp. 112–113 concludes that there is no evidence that Martin invented a nautical dromometer, citing Robert Heath, *Astronomia Accurata; or the Royal Astronomer and Navigator*, for the author, London, 1760, where a correspondent's scathing comments on Benjamin Martin's *New Principles of Geography and Navigation*, for the author, London, 1758, includes a disparaging reference, p. 297, to Martin's "*new Compass* and *Dromometer*", and on p. 312, "he [Martin] has contrived his *Compass*, which is to be 100 Times more exact than the common Sort, and his *New Nautical Dromometer*, by which he can get the *Course to a Minute*, and *the Distance to the utmost Exactness*." Did Grant mean to refer to the American-born loyalist, former merchant and surveyor, Benjamin Marston (1730–1792), who left Canada for London in 1787 seeking compensation for property lost during the War of Independence: for an outline biography see *Dictionary of Canadian Biography* – accessed at www.biographi.ca/en/bio/marston_benjamin_4E.html. Whilst in London Marston designed (1789) a navigational instrument to "determine the sun's altitude when the horizon is invisible": see Violet Mary-Ann Showers, *The Price of Loyalty: the Case of Benjamin Marston*, University of New Brunswick, MA Thesis, 1982, available on line at https://www.genealogy.com/ftm/s/t/y/Carl-W-Stymiest/FILE/0005page.html. Late in 1791 he approached the Board of Longitude with his proposal for "an instrument to measure a Ship's way at Sea, called a Marine Dromometer". The Board requested a full description and "proper Drawings": Cambridge University Library, RGO 14/45, fol. 243–244; these were not forthcoming, probably because in January 1792 the indigent designer accepted a post as surveyor to a company proposing development of an island entrepôt off the coast of West Africa. He left England in April and died of fever at the new settlement in August 1792. It is unlikely that Grant would have heard of Marston's design.

and the Board of Longitude in 1788, to no avail in both instances. However, returning to France with this and other instrumental inventions, he aroused sufficient interest to have the design published in 1803. A translation published in an English technical journal produced mentions in others, as if it were a new design.[21] Another example is the perpetual log patented by R. H. Gower in 1792, during a gap in service after completing his first senior posting as third officer of East Indian Company *Worcester* (1789–90). The specification was supplemented by a 32-page pamphlet describing the genesis and use of the instrument.[22] The pamphlet records Gower's admission that "experiment soon shewed me the difference between theory and practice". Constructive comments on his theoretical analysis came from John Adams (1738–1796?), a teacher of mathematics and navigation, whilst the final design was materially influenced by the advice from the London mathematical instrument maker Gabriel Wright. At the latter's recommendation, Gower worked with "an ingenious mechanician", Alexander Grant, "who undertook to construct the invention." Gilbert and Wright, the instrument makers to the East India Company, were described as the sole makers and vendors, but there is no evidence that the design went into commercial production.[23] A radical re-design was published in 1807 (*Fig. 6.2*), in part influenced by trials that Gower undertook on a canalised section of the River Lea. Here again there is no evidence that the design was produced commercially, though an example was used on Gower's experimental merchant vessel, *Transit*.[24]

---

21 D. J. Bryden, *Innovation in the Design of Scientific Instruments in the Georgian Era: The Role of the Society of Arts*, William Shipley Group for RSA History, Research Paper 3, 2019, p. 173. See also "Nouveau lock pour determiner la longitude en mer", *Annales des arts et manufactures* 13 (Paris 1803), pp. 310–312; reprinted in translation as "Description of a new log to ascertain longitude at sea by the aggregate of a ship's way. Compared with the ordinary log", *Repertory of Arts, Manufactures and Agriculture*, 2nd series, 5 (1804), pp. 127–128.

22 English patent 2350 (1792), "Perpetual log, or instrument for measuring a ship's way through the water, and for ascertaining the rate of sailing at any time". Richard Hall Gower, *A description of several instruments for measuring a ship's way through the water … with an account of his patent perpetual log*, for the author, London, 1792. Gower had gone to sea with the East India Company (EIC), as a midshipman, aged 13. For his EIC service as an officer, see Charles Hardy, *A Register of Ships employed in the service of the Honourable the United East India Company 1760–1819*, 3rd edn., revised and with additions by Horatio Charles A. Hardy, Black, Kingsbury, Parbury & Allen, London, 1820, pp. 141, 160, 174.

23 Not listed in *A Catalogue of Mathematical, Optical and Philosophical Instruments, made and sold by Gilbert, Wright and Hooke*, London, [1793]; nor mentioned in Richard Hall Gower, *A treatise on the theory and practice of seamanship*, G. G. & J. Robinson and Gilbert, Wright and Hooke; 2nd edn. corrected and enlarged, G. G. & J. Robinson, 1796; 3rd edn., Wilkie & Robinson *et al.*, London, 1808.

24 Richard Hall Gower, *Supplement to the Practical Seamanship*, J. Mawman, London, 1807, part II "Observations on the log and line with a description of various instruments for measuring a ship's way and an account of R H Gower's Patent-Log", pp. 97–141. There

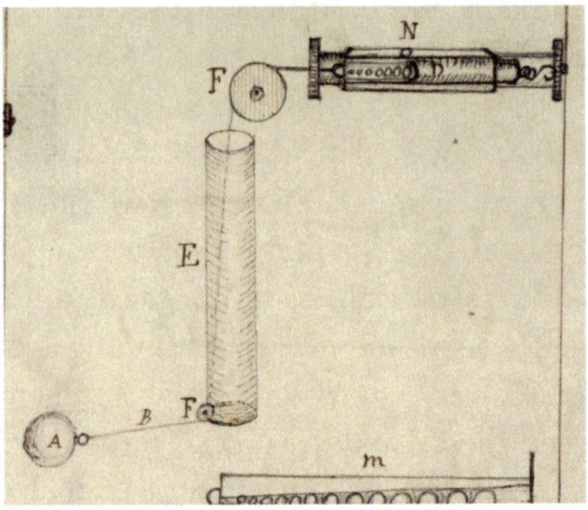

FIGURE 6.1
Charles Grant's perpetual log, or "Hydroscope", 1806
CAMBRIDGE UNIVERSITY LIBRARY, RGO 14/44, FOL. 277

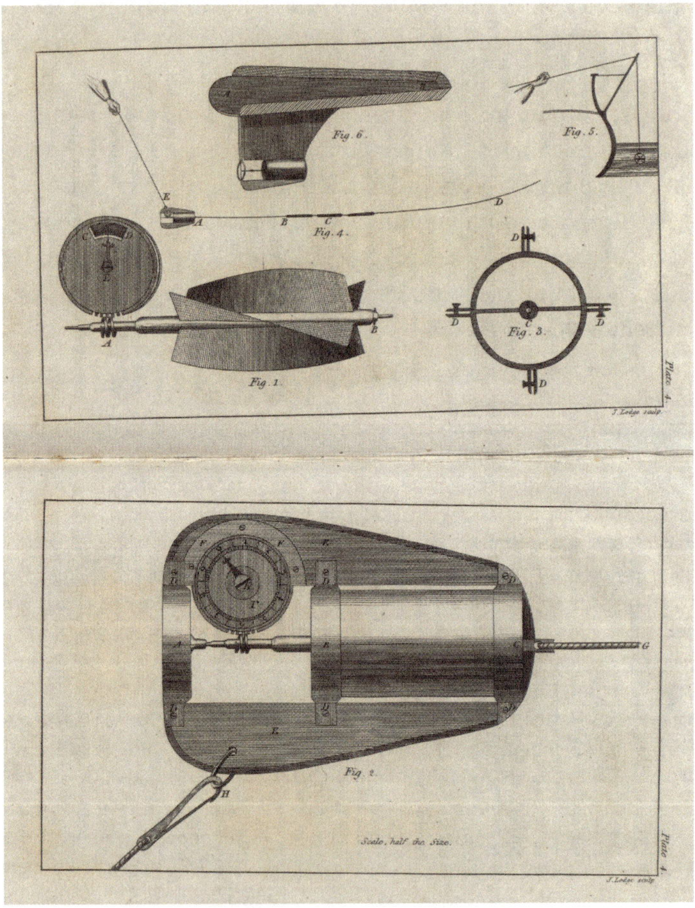

FIGURE 6.2 Richard Hall Gower's improved patent log, 1807, from Gower's *Supplement to the Practical Seamanship*, plate 4
NATIONAL LIBRARY OF SCOTLAND, REF. C.FRAS.51, REPRODUCED WITH PERMISSION

In the very early years of the reign of Victoria, John Washington RN, distinguished for his marine survey work, and later to become Admiralty Hydrographer, wrote a report on the perpetual log – 'Montre à sillage' – of the Rochefort watchmaker Leandre Clement, patented in England in 1838. Washington opened with a survey of "what has been done in former times as to finding a substitute for the common log, which, it must be confessed, is a sufficiently primitive method of measuring the speed of a ship". His initial entries were drawn directly from Grant, and from his naval confrères he added the designs of Bingham (1824) and Evans (*c.*1832). Of the Massey patent log (1802 and 1836) he states "too well known to need a description, and as far as our experience goes shews the distance run correctly".[25] The Massey log was added to the entry in Burney's 1815 edition of Falconer's *Dictionary* – but culled directly from a secondary account, and then left unchanged in the 1830 edition.[26] Not until 1839 and the twelfth edition of Norie's classic text book, was there inserted a footnote acknowledging alternatives to the log-line and half minute glass: "Several machines have been invented for measuring the distance run by a ship in a given time; but the best of these is *Massey's Patent Log*. This machine consists of two parts .... [brief description follows] .... This log has been in use several years, particularly in the Royal Navy."[27] Massey's patent log is discussed later in this paper, but despite the late insertion in Norie's text, it is noted that as early as 1816 Norie included the Massey patent log, priced at £5 5s with directions, in the list of nautical instruments he sold.[28]

---

is no evidence of Gower's log being made at the manufactory that the Gilbert Brothers established in the 1820s; see D. J. Bryden, "The 1828 bankruptcy of W. & T. Gilbert, Optical & Mathematical Instrument Makers of Leadenhall Street, London and Buckhurst Hill, Woodford, Essex", *London's Industrial Archaeology* 16 (2018), pp. 1–29.

25 John Washington, "M. Clement's Nautical Inventions: the Sillométre, Sub-Marine Thermometer, Steam Indicator, Derivométre, and External and Internal Thermometer", *Nautical Magazine* 12 (1843), pp. 372–374; the whole article reprinted in *Journal of the Franklin Institute*, 3rd series, 6 (1843), pp. 348–356.

26 Falconer/Burney, *op. cit.* (n. 9), (1815), pp. 237–238; (1830), pp. 238–239; drawing on "Description and use of the sea-log and sounding machine invented by Mr Edward Massey of Hanley, in Staffordshire", *Journal of Natural Philosophy, Chemistry and the Arts* 21 (1808), pp. 245–246.

27 Norie, *op. cit.* (n. 4), 12th edn. for the author, London, 1838, p. 59; that entry and footnote remained unchanged until at least the 20th edn. (1872).

28 J. W. Norie & Co., *Catalogue of the latest and most approved Charts, Pilots and Navigation Books*, Norie & Co., London, 1816, p. 12. Norie & Co. cited as London agents for the sale of the sounding machine and log in Massey's advertisement for his newly patented marine chronometers: *Liverpool Mercury*, 26 May 1815.

4 Trials

Many design proposals never got beyond the theoretical, but prototypes of others were subjected to sea trials. The Worcester physician Conyers Purshull had announced "a method proposed to find out the exact rate that a ship runs, and consequently the longitude at sea" in 1705, and he included that design, which in effect measured wind strength, as a means of ascertaining leeway, when he submitted a design for a perpetual mechanical log to the Commissioners of Longitude in 1715. His 1720 offer of ten per cent of any reward to the "Captain or Master of a Ship, or other Person, [who] will undertake to get this Engine made, as above described, and make the Experiment of it at Sea, so as to satisfy the Government of the Truth of its Performance",[29] does not appear to have born fruit. Guernsey resident Henry De Saumarez submitted his proposal for a perpetual mechanical log to the Royal Society in 1715. He had been unable to have an instrument made in Guernsey "for want of able Workmen", so came to London, and had a prototype made and tested on the Thames. The Royal Society indicated that the matter was not within their competence, that the idea was not new, and that the permanent trailing of the log would slow the vessel. Saumarez petitioned the King, who instructed the Admiralty to consider the proposal. The latter asked Newton's advice. He indicated that the instrument would not be as exact as the log-line, that engineer Thomas Savery had proposed such a device in about 1710, and that even if Saumarez were to patent the device it would have to be proved to be more accurate that the log-line; adding "I have no experience in Sea Affairs, nor ever was at Sea, and therefore my Opinion is not to be relied on, without the Opinion of Trinity House".[30] Trinity House met the inventor, examined the instrument, and drawings. Saumarez added to his petition his interview with the mariner who had tested the Savery instrument, and found it "cumbersome and

29 [Conyers Purshall], *An Essay at the Mechanism of the Macrocosm: or the dependance [sic] of effects upon their causes. In a new hypothesis, ... with a method proposed to find out the exact rate that a ship runs, and consequently the longitude at sea*, J. Wale, London, 1705, pp. 331–332; Conyers Purshull, *To the Lords Commissioners for the Longitude*, [London, ?1715]: copy in Yale, Beinecke Library: BrSides 1988 113; *London Gazette* 5906, 22–26 November 1720.

30 Henry de Saumarez, *An account of the proceedings of Henry de Saumarez, ... concerning his discovery of an invention, by which the course of a ship at sea may be better ascertained than by the log-line, viz, by a dial placed in some convenient place of the ship, with a little bell striking from time to time the geometrical paces, miles, and leagues which the ship hath run*, for the author, London, 1717, p. 8. In this booklet Saumarez reprints his successive petitions and the formal responses of the authorities to whom the design was submitted.

troublesome, … useless in the storm or rough sea". Trinity House raised objections to Saumarez's design, which the inventor countered. He also complained that the time they had taken to come to the conclusion that the proposal was "wholly impracticable" had provided opportunities for his intellectual property to be pirated. Furthermore, his enquiries relating to a patent had been rebuffed, awaiting the Admiralty's formal response, which in turn merely repeated the Trinity House conclusions. Saumarez finished his pamphlet by requesting that "his Invention may be Examined, Viewed, and Experimented, that by the Merit and Usefulness of it may no longer be stifled, but that it may be suitably Encouraged by the Government". In the event de Saumarez had to make the running. In the early 1720s he supplied examples to mariners working out of Guernsey, some to the Mediterranean, others to South America. All provided positive performance reports. His activities became known to several savants, most notably J. T. Desaguliers FRS, and since it was considered that the instrument might be affected by local currents, he was advised to test the design on still water. Permission was given to make trials on the canal in St. James's Park. The Royal Society were sufficiently impressed to publish a detailed account of the design,[31] and to recognise that his mechanical log could also be used to measure tidal flows.[32] Among those Saumarez made contact with was Samuel Molyneux FRS, who was appointed a Lord of the Admiralty in July 1727, and in that role he and other Commissioners observed small boat demonstrations on the lower reaches of the Thames. Molyneux arranged for Saumarez to test the design against the conventional log-line when the Royal Yacht *William & Mary* made a trip to Holland. Dutch observers on board were impressed, and the inventor went to The Hague to trial the instrument at different speeds on canals. Some design tweaks were suggested, but could not then be tried as Saumarez had to return to London on HMRY *William & Mary*. Molyneux died shortly after the ship returned home. The inventor lost his Admiralty patron, and the opportunity of further sea trials by the Navy.[33]

31 Henry de Saumarez, "An account of a new machine called the Marine Surveyor, contrived for the mensuration of the way of a ship in the sea more correctly than by the log; or any other method hitherto used for that purpose; together with several testimonials setting forth the usefulness of the invention", *Philosophical Transactions of the Royal Society* 33 (1725), pp. 411–432.

32 Henry de Saumarez, "Observations upon the Tides in the River Thames", *Philosophical Transactions of the Royal Society* 34 (1726), pp. 68–72.

33 Henry de Saumarez, "A further account of a new machine called the Marine Surveyor, designed for the mensuration of the way of a ship at sea, more correctly than by the log", *Philosophical Transactions of the Royal Society* 36 (1729), pp. 45–58.

Reports of sea trials indicate that a design had at least reached the prototype stage. This is the case for the designs of Reynoldson (1731),[34] Telfer (1744),[35] an unnamed gentleman of Derby (1767),[36] Walker (1786); two French inventors who brought their designs to England, Leguin (1787),[37] and Vallet (1789);[38] and Elvey (1830).[39] The expressions of approval from merchant mariners who had either seen or trialled Richard Benson Walker's prototype, impressed the Board of Longitude, whom he approached in 1790. The Admiralty was asked to conduct a trial, with the Board agreeing to pay Walker's costs in making an

---

34 English patent 534 (1731), "Machine that gives exactly the way a ship makes, and counts her leeway, tells what speed the winds blow, and proves the different force of current in any depth". A positive report was read at a meeting of the Royal Society late in 1732; see Royal Society Archives: RBO.18.11 for report of Reynoldson's "hydrographical machine for measuring a Ships way at Sea" on a run from Bell Dock Wapping to Gravesend on the yacht *Prince William*.

35 *Newcastle Courant*, 17 March 1744 citing information from Edinburgh, records positive results of a "trial of a machine invented by Mr Telfer, Watch-maker in Glasgow, for measuring a Ship's Way on the sea". The trial was held in the Firth of Forth over distances of 5¾ and 1 mile, as taken from charts. Nothing further is known, but the impact of the 1745 Jacobite Rebellion, and Telfer's drowning at sea on a passage from Dublin to Irvine may account for that: see *Caledonian Mercury* (Edinburgh), 23 December 1746.

36 *Manchester Mercury*, 9 June 1767; *Public Advertiser* [London], 2 July 1767; *Derby Mercury*, 10 July 1767; "Useful Projects", *Annual Register [for 1767]* 10 (1768), p. [103], all report the "new marine surveyor" being tested at sea "off the Downs", a location implying use of a naval vessel. The success or otherwise of the trial were not reported – presumably it failed to meet expectations. The designer was the entrepreneurial cartographer Peter Perez Burdett, who lived in Derby in the 1760s before moving to Liverpool: see *Oxford Dictionary of National Biography* https://doi.org/10.1093/ref:odnb/37243. The mariner William Hutchinson, subsequently Liverpool Dock Master, who had tried Smeaton's sea way measurer at sea (see n. 130) commented "Mr. *Burdet*'s contrivance seems to me a further improvement [on Smeaton's design], which I think deserves more trials and encouragement from those in power": see William Hutchinson, *A Treatise on Practical Seamanship*, for the author, Liverpool, 1777, p. 109.

37 Submitting his design to the Society of Arts, Leguin indicated that he had trialled and calibrated the instrument on a voyage between Calais and Brest, but he had left the details at Brest: see Bryden *op. cit.* (n. 21), p. 173.

38 Vallet, with the 1782 closure of the bleaching fluid manufactory that had brought him to England, submitted his 1789 designs for ships logs to the French authorities (1790) and the Society of Arts (1796). He reported trials on voyages from Liverpool to Dublin. However, his personal circumstances were such that the Society of Arts was unable to make contact with him, so they made no judgement: "if the machine had succeeded the Inventor would have met with suitable reward": Bryden, *op. cit.* (n. 21), pp. 221–222.

39 John Elvey was an inventive Canterbury millwright, and patentee: for the genesis and sea trial of his instrument to show a ship's way and leeway, see *Kentish Chronicle*, 19 January 1830 and 11 May 1830, 31 May 1831 and 25 October 1831; also *Maidstone Gazette and Kentish Courier*, 17 January 1832.

instrument and installing it. In the event, the commander of the sloop HMS *Hound* reported that it was "very liable to be out of order and had stopped on his passage from the Downs to Spithead". Walker informed the Board that he knew how to solve the problem, but after promptly paying his expenses they let the matter rest.[40]

Serving officers, like Owen (*c.*1800),[41] Hamilton (1801),[42] Bingham (*c.*1820),[43] Graves (*c.*1830),[44] and Evans (*c.*1832),[45] were in a position to have their designs tested at sea without necessarily requiring formal Admiralty approval. In contrast the side-lined James Burney (1809), on half pay since 1782, had to be

40 Cambridge University Library [CUL]: RGO 14/45, fols. 339–343; CUL: RGO 14/6 Minutes of the Board of Longitude, 6 March 1790 and 4 December 1790. See also TNA: HO 42/17/7, fols. 13–14 for Walker's letter to the King reporting the success of trials. Walker, a merchant of Kingsland, Middlesex, was frequently financially stretched, being bankrupt more than once: *London Gazette*, 7 December 1784, 26 September 1795, 23 December 1806.

41 Mackay, *op. cit.* (n. 11), pp. 10–11 – "lately communicated to the author" – so post 1797 when William Fitzwilliam Owen was appointed Lieutenant and prior to 1803/4 when he sailed for the East Indies.

42 Thomas Hamilton, "Principle for a contrivance to measure a ship's rate of sailing, and to determine the quantity sailed in a given time", *Naval Chronicle* 6 (1801), pp. 125–131; the reprinted version in *Repertory of Arts, Manufactures and Agriculture*, 2nd series, 1 (1802), pp. 355–364, erroneously describes Hamilton as a Captain in the Royal Navy – he was a civilian volunteer, in command of the Sea Fencibles at Margate. In that role he was able to invite a naval Captain, a Lieutenant and the Master of HMS *Isis* to observe trials, presumably on his own merchant vessel. A brief mention of the design in *Retrospect of Philosophical, Mechanical, Chemical and Agricultural Discoveries* 2 (1806), p. 318, conflates the identification error, by attributing it to "Captain Hamilton (now Commissioner of the Navy)". The *Repertory* version adds that the Admiralty had subsequently authorised trials on HMS *Argo*.

43 Noted by Washington, *op. cit.* (n. 25), p. 273, as "fitted to the keel of the *Tourist* steamer, was somewhat similar to the *Navivium* of Gilmore as far as we can learn." Earlier, as Flag Captain on HMS *Dover*, based at Leith, Arthur Batt Bingham had trialled the design, forwarding performance reports to the Board of Longitude, who dismissed it: "the Board could not discover any material improvement in principle that could sufficiently distinguish it from former inventions of the same kind to justify them in declaring their opinion in its favour.": CUL: RGO 14/7, Board of Longitude Minutes, 5 February 1824; see also CUL: RGO 14/31, fols. 246–249.

44 Thomas Graves, "An improved log-ship", *United Service Journal and Naval and Military Magazine* (1831) part 2, pp. 501–502: "its construction has been approved of by many naval officers, several of whom have had opportunities of using it; and from repeated trials of my own, it answered wall the objects proposed". Graves joined the navy 1816, was commissioned 1827 and appointed to the surveying vessel HMS *Adventure*, serving on surveying vessels until 1846: see William R. O'Byrne, *A Naval Biographical Dictionary*, John Murray, London, 1849, 3 vols.: v. 1, p. 424.

45 Washington, op. cit. (n. 25), p. 373: fitted to the steamer HMS *Rhadamanthus* "but was found not to answer".

content with having his proposal published.[46] Russell (*c.*1750), highly placed in the Levant Company, eventually had an evolved design tested on the Admiralty polar expedition of 1773, as did Foxon (1772).[47] Gottlieb (1791) had his perpetual log trialled by the government on a cutter, with sufficient success for the design to be fitted to the first two of a new fleet of fast Post Office mail boats.[48] Hookey (1827), a naval midshipman was awarded a silver medal by the Society of Arts in 1827 for his fish-shaped log (*Fig. 6.3*), with the key evidence being the positive report from the master of a naval supply vessel of performance on a Deptford to Leith voyage.[49] Francis Higginson, mate RN, submitted his spring loaded log and instrument to measure leeway to the Society of Arts and the Board of Longitude in 1828 (*Fig. 6.4*). The Admiralty had sanctioned trials which were carried on the cutter HMS *Arrow*. The Society of Arts was critical of methodology and conduct of the trials, and at the same time considered the design was far from original. The Board of Longitude, about to be wound-up, made no judgement.[50]

Baines (1816), partner in a Hull glue factory, patented his perpetual log (*Fig. 6.5*), and the design submitted to the Board of Longitude was accompanied by positive letters of commendation from merchantmen and naval survey vessels who had used it. The Navy Board instructed that one be trialled on a seagoing vessel, possibly the brig sloop HMS *Cameleon*.[51] One was certainly fitted to the specially converted Hull-built whaler HMS *Isabella*, one of the

---

46 James Burney, "New method proposed for measuring a ship's rate of sailing", *Journal of Natural Philosophy, Chemistry and the Arts* 24 (1809), pp. 57–8; abstract in *Retrospect of Philosophical, Mechanical, Chemical and Agricultural Discoveries* 5 (1810), p. 347, with the editorial observation "the whole of what induced Mr. Burney to propose this new method seems to have been only guessed work, and therefore not at all to be depended upon". Despite a history of ill discipline, Burney was commissioned Captain 1782; further insubordination led to retirement on half-pay from 1784. Subsequent petitions for re-employment were rebuffed: see the entry in *ODNB* https://doi.org/10.1093/ref:odnb/4080.

47 See above, Phipps, *op. cit.* (n. 15).

48 Bryden, *op. cit.* (n. 17).

49 Bryden, *op. cit.* (n. 21), p. 162: the prototype survives: Science Museum (London), inv. no. 1920–374, donated by the Admiralty.

50 Bryden, *op. cit.* (n. 21), p. 156: an example is extant: Royal Museums Greenwich, NAV0730 – no provenance recorded.

51 English patent 4082 (1816), "Perpetual log or sea perambulator": the specification was printed in *Repertory of Arts, Manufactures and Agriculture*, 2nd series, 37 (1820), pp. 65–68; a non-technical synopsis appeared in *New Monthly Magazine and Universal Register* 14 (1820), p. 214; reprinted in *The New York Literary Journal and Belles Lettres Repository* 3 (1820), pp. 469–470. See also CUL: RGO 14–31, fols. 225–227 for the handbill headed: "*TO SHIP-OWNERS. R. Raines Baines, BEGS leave to offer to SHIP OWNERS, COMMANDERS &c &c his Sea Perambulator, Or, PERPETUAL LOG.*" Use by HMS *Cameleon* is reported by Baines in his letter to the Secretary of the Board of Longitude, 25 May 1819 – CUL: RGO 14–31, fol. 232.

FIGURE 6.3 James Hookey's fish-shaped log
SCIENCE MUSEUM, LONDON, 1920–374, PRESENTED BY THE LORDS COMMISSIONERS OF THE ADMIRALTY

FIGURE 6.4 Francis Higginson's spring-loaded log, circa 1828, with detail shown below
NATIONAL MARITIME MUSEUM, GREENWICH, NAV0730. PHOTO © NATIONAL MARITIME MUSEUM, GREENWICH

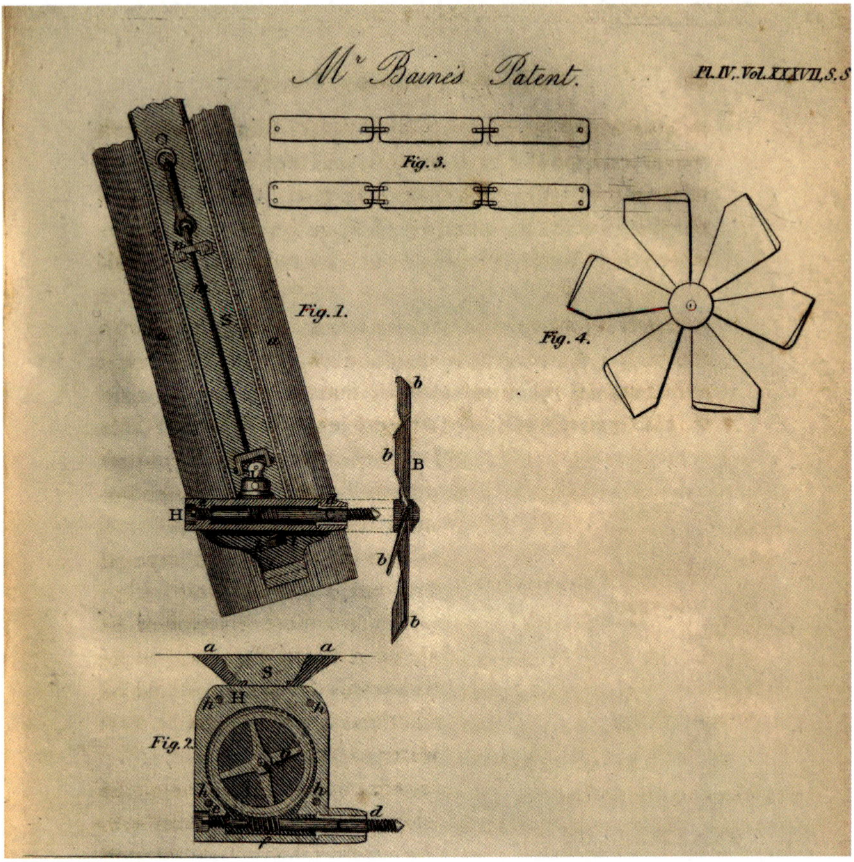

FIGURE 6.5 R. R. Baines's perpetual log or "sea perambulator", 1816. From the *Repertory of Arts, Manufactures, and Agriculture*, 2nd series, 37, 1820; plate altered to remove extraneous figures
COURTESY OF THE UNIVERSITY OF MINNESOTA

Ross expedition ships, and evaluated with other scientific apparatus on the 1818 Admiralty-funded expedition to seek the North-West Passage. Despite the early failure of a critical part, Ross was positive about the design, "this instrument would be of great use, particularly to surveying vessels, as it is capable of measuring a distance with great accuracy".[52] The Board of Longitude, in

52 John Ross, *A voyage of discovery, made under the orders of the Admiralty, in his Majesty's ships Isabella and Alexander, for the purpose of exploring Baffin's Bay, and inquiring into the probability of a north-west passage*, John Murray, London, 1819, pp. xvii–xii lists the instruments taken on the voyage, with their performances summarised: appendix pp. cxxiv *et seq.*; p. cxxxi for *Baines Log*. The quotation is from the 2nd edn., Longman, Hurst, Rees, Orme & Brown, London, 1819, 2 vols.: v. 2, p. 228.

contrast, considered that there was no novelty in the design principle, nor any decided superiority in the construction.[53] Trade directory entries in the 1820s describing Baines as a "patent log manufacturer" suggest that the dismissal of the design by the Board of Longitude does not appear to have persuaded him to abandon commercial production of the design.[54]

Jean Leandre Clement (1838), then a watchmaker, based in Rochefort, came to England to patent and press the case for his designs with the Admiralty. English periodicals reported that the performance of the Clement log in trials on a French naval steam vessel had been such that it had been ordered to be fitted to their powered naval vessels.[55] Perhaps it was this claim that stimulated interest by the Admiralty. Both Clement's on-board log (which he named in English a "sillometer" from the French "montre à sillage"), his submarine thermometer, his current meter, and his engine indicator were given exhaustive British naval trials. The first of these took place in October 1842 on board HM Steam Vessel *Lightning* in cruises in the Medway and on a voyage from

---

53 CUL: RGO 14–7, Minutes of the Board of Longitude, 1 April 1819 and 3 June 1819.

54 Edward Baines, *History, Directory and Gazetteer of the County of York*, Edward Baines, Leeds, 1823, 2 vols.: v. 2, pp. 263, 300, 329, 336, lists Baines as a partner in Porter Raines & Co., glue manufacturers, and on his own as a baker; *Pigot and Co.'s National Commercial Directory for 1828–9*, J. Pigot & Co., Manchester, 1828, pp. 972, 985, list Baines as a baker and flour dealer, and as a patent log manufacturer. He died, aged 50, in 1828: see *Hull Packet*, 3 June 1828.

55 English patent 7780 (1838), "Ascertaining and indicating the rate of vessels passing through water"; *Inventor's Advocate and Journal of Industry*, no. 108, 21 August 1841, pp. 121–122; *Mechanics Magazine*, 37 (1842), p. 34. The 15-year French patent for the "montre à sillage" was granted 23 October 1838: see *Catalogue des Brevets d'Invention, d'importation et de perfectionnement … 1828–1842*, V. Bouchard-Huzard, Paris, 1843, p. 100. The positive performance of Clement's "montre à sillage" had been first reported to the local improvement society in November 1837, and all three of his marine instruments were the subject of a detailed report 10 months later, following five months of trials on the French naval steamer *Le Grégeois:* see Jean-Pierre Dinand, "Pendant le Monarchie de Juillet (1830–1845), … Le Société d'Agriculture, Sciences et Belles-Lettres (1834–1848)" in *Les Sociétés Savantes de Rochefort (1806–2006)*, Société de Geographie de Rochefort, Rochefort, 2007, p. 48 and *Société de Littérature, Sciences et Arts … de Rochefort: Procès-verbaux des séances … 1830–1847*, Société de Geographie de Rochefort, Rochefort, 2006, pp. 169, 174. The sillometer (montre à sillage) was early reported in the French press: *Le France industrielle, manufacturière, agricole, artistique*, 7 March 1839; *Le Censeur, Journal de Lyon, politique, industriel et littéraire*, 20/21 March 1839; and in Germany, *Blätter für literarische Unterhaltung* (Leipzig), 5 May 1839 and *Das Pfennig Magazine der Gesellschaft* (Leipzig), 18 April 1840. *Le Magasin Pittoresque* 2 (Paris, 1842), p. 159 reports "la montre à sillage, que l'on installe dans ce moment à bord de plusiers navires, par ordre du minister de la marine."

Woolwich to Portsmouth and return.[56] This and subsequent trials were presumably with instruments incorporating the improvements of his 1842 patent. Clement may have fed the press with accounts of this trial: "so satisfactory, that the Lords Commissioners of the Admiralty have directed that £200 be paid to M. Clement, and the machinery deposited in Woolwich dockyard".[57] There was a second trial in April 1843 in which sillometer, submarine thermometer and engine indicator made to Clement's designs by the London instrument maker Cary, were fitted at Woolwich to HMSV *Blazer* and rigorously tested on a voyage from Woolwich to the Nore. Clement, together with Cary, and the member of the dockyard staff who had supervised their installation, were all in attendance.[58] The frigate HMS *Penelope*, being converted to operate as a paddle steamer was fitted with a Clement sillometer, and made her first passage in September 1843.[59] But that appears to close official British interest in Clement's designs. The compiler of the official report, Captain John Washington RN, was sufficiently impressed by the performance of the sillometer to refer to it, together with Massey's log, as being sufficiently sensitive to measure submarine currents.[60]

The activities of the Canterbury millwright and patentee, John Elvey, indicate the problems of a designer lacking the right contacts. Elvey's idea for a device to show a ship's way and leeway "without heaving the log" (1830) was to have been brought to the attention of the Admiralty by a local aristocrat, but the latter's death left him without patronage and financial support to take the

56 "Mr Clement's Nautical Inventions, the sillometer, marine thermometer and steam thermometer", *Mechanics' Magazine* 37 (1842), pp. 410–413 gives a full account of the trials. This report reprinted in *Civil Engineer and Architect's Journal* 6 (1843), pp. 28–29.

57 *London Standard*, 11 November 1842; *Morning Post* (London), 11 November 1842; *Athenaeum*, no. 785, 12 November 1842, p. 974. Fuller technical data is provided in *Nautical Magazine*, 12 (1843), pp. 330–336. The story was picked-up and reprinted widely, not only in British newspapers and magazine, but also in the United States; e.g. *Eclectic Museum of Foreign Literature, Science and Art* (Philadelphia), united series, 1 (1843), p. 136.

58 Washington, *op. cit.* (n. 25). For Cary's 1843 description of sillometer, marine thermometer, and steam thermometer see Hydrographic Office Archive, Taunton: MLP 28/1/1.

59 *Morning Post*, 14 September 1843; *Morning Chronicle*, 14 September 1843. The conversion of *Penelope* from sail to steam attracted considerable interest, see *Morning Post*, 30 June 1843; *Nautical Magazine and Naval Chronicle*, 8 (1843), pp. 568–569; *Illustrated London News*, 1 July 1843. It was not universally judged to be a success: *Hampshire Advertiser*, 14 December 1844, citing *Plymouth Times*.

60 John Washington, "Report on the quantity of silt held in suspension by the steam in Dover Bay in February 1845", in "First Report of the Commissioners … on the state of harbours, shores and navigable rivers", *Parliamentary Papers, Reports from Commissioners* 14 (London 1845), appendix 38, p. 197. Washington served as Hydrographer to the Navy 1855–1863.

project forward. Nevertheless, the proposal was developed. Subject to a successful sea trial, "a gentleman of high rank in the navy" indicated willingness to present the design to the Admiralty. However, the sea trial between Whitstable and London, indicated the need for further design adjustments,[61] and there the matter appears to have ended.[62] The inventive entrepreneur Mark Cosnahan, a Manx merchant and banker, patented his "marine computist" in 1825.[63] He had earlier outlined what the device would achieve, but not how.[64] Though he was subsequently (1826–1829) the owner of the auxiliary paddle steamer *Victory*,[65] Cosnahan may not have bothered to take his design to the stage of testing a prototype at sea; perhaps he was discouraged by claims of unoriginality made when abstracts of his patent were published.[66]

For some designers, having a prototype of their proposal made and tested, whether in a lake, canal, navigable river, estuary or at sea, might on its own have constituted a measure of success. Even a design that got no further than a proposal in principle could be considered a success if publication added to the author's contemporary status. Henri Pitot's 1732 method of measuring water flow, which he noted could also be used to give a ship's speed, was widely recognised as important by savants,[67] though the application to water flow

---

61 *Kentish Weekly Post*, 19 January and 11 May 1830; 31 May and 25 October 1831; *South Eastern Gazette*, 17 January 1832.

62 Elvey's patent winnowing machine, English patent 3512 (1811), was marketed: *Kentish Weekly Post*, 22, 25, and 29 October 1811. Despite financial set-backs, he continued to invent, taking out further patents: D. J. Bryden, "John Elvey (1776/7–1844): Inventive Canterbury Millwright", Mills Archive 2023 available at https://catalogue.millsarchive.org/john-elvey-1776-7-1844-inventive-canterbury-millwright.

63 English patent 5130 (1825), "Apparatus for ascertaining the way or leeway of ships and other vessels".

64 Mark Cosnahan, *Important observations for the benefit of commerce and navigation*, for the author, Liverpool, 1823, pp. 8–10.

65 Ernest Henry Steening, "Records of the Cosnahan Family", *Proceedings of the Isle of Man Natural History and Antiquarian Society* 4 (1942), pp. 516–526. *The Isle of Man Steam Packet Co., Ltd. Centenary 1830–1930*, Isle of Man Steam Packet Co., Douglas, 1930, p. 14.

66 "Account of new patents", *Repertory of Patent Inventions and other discoveries and improvements in Arts, Manufactures and Agriculture*, 3rd series, 2 (1826), pp. 394–396; "Recent patents", *London Journal of Arts and Sciences* 13 (1827), pp. 18–20.

67 Henri Pitot, "Description d'une Machine pour mesurer la vitesse des Eaux courantes et le sillage des Vaisseaux", *Histoire de l'Académie royale des sciences* 34 (1735), pp. 363–376. The use in measuring a ship's speed is considered towards the end of the article, where a trial on a small boat on the Seine is described. More widely known through inclusion in subsequent editions of Bernard Forest de Bélidor, *L'architecture hydraulique, ou l'art de conduire, d'elever, et de menager les eaux*, C. A. Jombert, Paris, 1737, 4 vols.: v. 1, pp. 255–256.

measurement did not become really effective until well into the nineteenth century,[68] and its use as the basis of speed measurement in aircraft occurred after almost two hundred years.[69] Nevertheless, Cooke (1792),[70] Hamilton (1801),[71] Hamill (1806),[72] and Evans (*c.*1832),[73] all proposed Pitot-tube type instruments to measure a ship's speed, with the latter conducting an unsuccessful trial.

5 Successful Designs?

The question mark is intentional. During the Hanoverian era, the log-line method of finding the way at sea, remained at the core of day-to-day navigational practice on British ships, both in the naval and the merchant service. Just one design, that of Massey (1802), broke the stranglehold, though those of Foxon (1772),[74] Gould (1800)[75] and Baines (1816)[76] briefly made inroads into the conventional operation. Foxon, who had served as a ship's carpenter for the East India Company, had evolved and tested his design over three passages to India. The taking-out of the patent may have been the stimulus

68 Glenn Owen Brown, "Henry Darcy's perfection of the Pitot tube", in G. O. Brown *et al.* (eds.), *Henry P. G. Darcy and other Pioneers in Hydraulics*, American Society of Civil Engineers, Reston, Virginia, 2003, pp. 14–23.

69 John Kirkham Bradley, "The History and Development of Aircraft Instruments – 1909–1919", University of London PhD Thesis, 1994, pp. 41–56. John K. Bradley, "Putting the Wind up the Pilot: Cloud Flying with Early Aircraft Instruments", *History of Technology* 18 (2016), pp. 95–112: 99.

70 CUL: RGO 14/6, Minutes of Board of Longitude 2 March 1793: Maskelyne reported "The Idea is ingenious, but it would require experiments in still Water, as well as at Sea, to shew whether it is practicable to a good purpose, notwithstanding the obvious objections of the motions of the Ship, and Waves of the Ocean."

71 This device was given naval trials; see above note 42.

72 Hugh Hamill, jun., "On the measure of a ship's velocity", *Journal of Natural Philosophy, Chemistry and the Arts* 14 (1806), pp. 343–344. Francis Hopkinson's second version of an instrument to measure a ship's way (1790), using a Pitot-tube design, *Transactions of the American Philosophical Society* 3 (1793), pp. 239–241, was widely reported in England; *Repertory of Arts and Manufactures* 1 (1794), pp. 49–52; *Memoirs of Science and the Arts* 1 (1794), pp. 129–130; *Monthly Review or Literary Journal* 14 (1795), p. 196; *Glasgow Mechanics' Magazine* 2 (1825), pp. 21–22.

73 Washington, *op. cit.* (n. 25) dates to 1830, but the paddle steamer HMS *Rhadamanthus* was not launched until 1832, with Evans responsible for fitting out, and becoming her Commander 1832–1835: see O'Byrne, *op. cit.* (n. 44), p. 341.

74 English patent 1028 (1772), "Machine for Measuring a ship's way".

75 See below and notes 86–91.

76 See above and notes 51–54.

for the Admiralty to order sea trials – including that undertaken by Phipps in 1773.[77] There appears to have been some attempts at "news management", for "success" of the Phipps's trial of Foxon's log was early and widely claimed in press reports.[78] It was not the sole trial, for six logs were delivered to the naval stores at Deptford, and a further twelve in late 1777,[79] with records indicating that on more than six occasions Foxon logs were issued to particular vessels.[80] The naval explorer James Cook, who at best could only have been aware of Phipps's reservations at second hand, tried one of Foxon's logs on the return leg of his second circumnavigation, borrowed or purchased at St. Helena from the captain of an East India Company vessel. Neither Cook nor his officers were particularly impressed.[81] For a few years, Foxon supplied the Navy with spares of the operative "spiral worm", which was trailed over-board – and so potentially subject to damage. Manufacture of the on-board recording mechanism was undertaken by the watchmaker William Lovelace. When he died in 1777 the Lovelace executors offered for sale forty of these machines.[82] A few Foxon count mechanisms survive, all carry a serial number. "157" is the highest recorded.[83] The complete apparatus appears to have retailed at six guineas.[84]

---

77 *London Magazine or Gentleman's Monthly Intelligencer* 41 (1772), p. 603, for a report dated 15 December 1772; see also Phipps, *op. cit.* (n. 15), pp. 97–8.

78 *Pennsylvania Packet* (Philadelphia), 16 August 1773; *Rhind's Virginia Gazette* (Williamsburg), 2 September 1773; *Lloyds Evening Post* (London), 8 December 1773. See also *Caledonian Mercury* (Edinburgh), 8 December 1773; *Derby Mercury*, 24 December 1773; *Town and Country Magazine* 5 (1773), p. 671.

79 TNA, at Royal Museum Greenwich, NMM: ADM 354/188/94; TNA: ADM106/1239/265 and 106/1239/ 281.

80 TNA: ADM 106/1223/35, ADM 1227/102, ADM 1229/100, ADM 1239/412, ADM 354/195/115, ADM 106/1261/141. There appears to have been a further trial instigated by a particular Captain in 1788: see TNA: ADM 106/1296/337.

81 J. C. Beaglehole (ed.), *The Journals of Captain James Cook on His Voyages of Discovery*, Hakluyt Society, Extra series, no. 35, Cambridge, 1961, 5 vols.: v. 2, pp. 665–666. Michael Edward Hoare (ed.), *The Resolution Journal of Johann Reinhold Forster, 1772–1775*, 4 vols.: v. 4, Hakluyt Society, London, 1982, p. 751.

82 *Morning Chronicle and London Advertiser*, 16, 20, and 21 May 1777.

83 Number 43 is in a private collection, illustrated in Harriet Wynter and Anthony J. Turner, *Scientific Instruments*, Studio Vista, London, 1975, pp. 88–89; number 71 is the Science Museum, London, inv. no. 1857–51; number 157 was sold at auction by Christies (London), 6 November 1997, sale 7784, lot 281.

84 That was the price quoted to the surveyor William de Brahm in the early 1770s: see Louis De Vorsey Jr., "William De Brahm's 'Continuation of the Atlantic Pilot', an empirically supported eighteenth century model of north Atlantic surface circulation", in Mary Sears and Daniel Merriman (eds.), *Oceanography, the past, Proceedings of the Third International Congress on the History of Oceanography ... 1980*, Springer, New York, 1980, pp. 718–733: 730.

The Foxon log achieved a degree of recognition – witness the response of the Society of Arts to the log submitted to them by Estienne Leguin in 1787: they refused to countenance a bounty, one reason for refusal being that "the principle of its motion is very similar to that of Mr Foxon now in considerable use".[85]

A design that certainly went into production was that of Chester Gould, patented in the United States in 1799 and in England the following year, when the designer returned to the country of his birth.[86] Two months after he had taken out the American patent, the instrument was being offered for sale, by prior subscription, in New York.[87] By April 1800 another New York newspaper recorded where the instrument could be purchased, and reprinted a positive letter of commendation from the sailing master of the naval frigate *United States* who had tested the prototype of "Gould's Patent Sea Log" on a voyage to Europe.[88] In the United States Gould's brother arranged for the printing of an eight-page pamphlet, indicating that the logs were on sale and giving directions for use. The text was largely letters of commendation from masters in the merchant marine, seven of whom had used the device on transatlantic voyages with five making shorter voyages on the eastern seaboard. In addition

85 Bryden, *op. cit.* (n. 21), p. 173.

86 Edmund Burke (ed.), *A List of Patents for inventions and design issued by the United States … from 1790–1847*, Washington DC, 1847, p. 171. A fire in 1836 destroyed most of the early records of the US Patent Office: for the probable text of the specification, see "Valuable Nautical Invention" in *Aurora General Advertiser* (Philadelphia), 10 June 1799, where the inventors are named as Chester and Calvin Gould; and *Gazette of the United States and Philadelphia Daily Advertiser*, 11 June 1799. English patent 2405 (1800) "Instrument or log for ascertaining a ship's distance at sea"; see also "Specification of the patent … for an instrument or log for ascertaining a ship's distance at sea", *Repertory of Arts and Manufactures*, 13 (1800), pp. 225–230. The specification reprinted in John Wilkes (ed.), *Encyclopaedia Londinensis*, for the proprietors, London, 1810–1829, 22 vols.: v. 12 (1814), pp. 882–883; see also "[Abstract of the patent]", *Monthly Magazine or British Register* 10 (1800), pp. 360–361; the specification reprinted in translation as "Beschreiben einer neuen Art von Log, oder eines Fahrtmessers fur Seeschiffe", *Annalen der Physik* 8 (1801), pp. 474–479; the editor of this journal was at pains to record the priority of Rheinhard Woltmann of Hamburg for a mechanical log, citing a 1790 publication, unaware of the various designs of mechanical logs that had been patented in England well before that date.

87 *New York Gazette and General Advertiser*, 23 July 1799; reprinted in Rita G. Gottesman, *The Arts and Crafts in New York, 1777–1799*, *Collections of the New York Historical Society* 81 (1954), p. 217.

88 *Commercial Advertiser* (New York), 26 April 1800; text from R. G. Gottesman, *The Arts and Crafts in New York 1800–1804: Advertisements and News Items from New York*, *Collections of the New York Historical Society* 82 (1965), p. 245. The text of this letter is reprinted in other newspaper advertising, for example by the bookseller William R. Wilder of Newport, Rhode Island, who offered examples for sale; see *Newport Mercury*, 6 May 1800, and repeated in later numbers that month.

to reprinting the positive comments of the sailing master of the naval frigate *United States* noted above, there was a favourable performance report from the commander of another frigate, *Boston*.[89] A fuller account of the design with some technical detail appeared in the first issue of a New York quarterly in the spring of 1801, which also printed the *United States* trial report.[90]

The second English patent contained a major design change.[91] Gould certainly had examples made to this specification manufactured for him in London. It was advertised in Philadelphia in late May 1801.[92] In June 1801 a letter of approbation was published in a New York newspaper in June 1801, from S. B. Davies, of the merchant vessel *General Washington* who had used a model when returning from Liverpool.[93] That letter was reprinted in a hand bill printed in Boston headed: "*Directions for Using* GOULD'S NAUTICAL PERAMBULATOR", which also included another positive report from a naval frigate, the *Philadelphia*, whilst ship master, W. Johnson of Charleston, indicated that he had purchased six machines from Chester Gould in London.[94] Gould's brother Horace presented an example to the Boston Marine Society in August 1801.[95] The device was advertised in Boston in September 1801, the

89 *Gould's Patent Sea Logs*, [Philadelphia?] 1801: copy in Hagley Museum and Library, call mark *VK581. G69. The dated letters of commendation are 1800, except for one dated 6 January 1801. A measure of the impact of Gould's publicity is that in October 1800 English papers picked up a report from the French press on the satisfactory performance of the Gould log when tested on the US frigate *Constitution*; see for example *Kentish Weekly Post*, 24 October 1800 and *Chester Courant*, 28 October 1800.

90 "Perpetual Sea-Log", *American Review and Literary Journal* 1 (New York, 1801), pp. 131–132, reprinted without acknowledgement of source in the September issue of *Union Magazine and Imperial Register* (London) 2 (1801), p. 196.

91 English patent 2458 (1800), "An improved instrument or log for ascertaining a ship's distance at sea". "Specification of the patent … for improvements on his new instrument or log for ascertaining a ship's distance at sea", *Repertory of Arts and Manufactures* 15 (1801), pp. 227–237.

92 *Philadelphia Gazette*, 29 May 1801: cited in Frank A. Kafker, "Notable Encyclopedias of the Late Eighteenth Century: Eleven Successors of the Encyclopédie", *Studies on Voltaire and the Eighteenth Century* 315 (1994), p. 241, n. 91.

93 *American Citizen and General Advertiser* (New York), 1 June 1801: Gottesman, *op. cit.* (n. 88), pp. 245–246.

94 *Directions for using Gould's Nautical Perambulator*, Boston 1801: Library of Congress, printed ephemera collection, portfolio 47, folder 11. In England, the term "perambulator" had been a synonym for the surveying wheel or waywiser throughout the eighteenth century. It may have been used in this instance from an awareness of the United States patent granted to the inventive serial patentee, Amos Whitmore of Massachusetts in September 1796 for a "Perambulator for measuring a ship's way".

95 Nath'l Spooner (ed.), *Gleanings from the records of the Boston Marine Society through its first century*, Boston Marine Society, Boston, 1879, p. 63.

advertisement also reprinted Davies's commendatory letter, together with that of Barnabas C. Knox, the commander of the *Merchant* – who had purchased his new improved "nautical perambulator" in London from Chester Gould and after his return across the Atlantic wrote to Gould's brother to report its excellent performance.[96] There was additional awareness of the second design in the United States through its appearance as an appendix item in the final supplementary volume of Thomas Dobson's 1803 American edition of the third edition of the *Encyclopaedia Britannica*, but this may represent no more than the publisher having available the specification as it had appeared in the *Repertory*.[97]

There is a surviving example of the on-board count device made to the second Gould English patent (*Fig. 6.6*) preserved in the Peabody Essex Museum at Salem. It carries a silvered plaque engraved: "*Goulds Patent* | BOSTON | N° 17".[98] I judge this to be an import from England, and speculate that the plaque covers a signature recording London manufacture and probably the phrase "By the King's Patent" – a sentiment that would not have been appreciated in Boston! In contrast to his success in getting American vessels to trial both versions of the device, and the positive use of letters of commendation used in marketing the design on the eastern seaboard of North America, I find no similar evidence for England. It would seem likely that whilst Gould was able to cope with bureaucracy of the English patent system, he was unable to interest London maritime interests in the design. Furthermore, there is no evidence that the initial interest in the United States was maintained. Gould himself moved on to take out patents in quite different fields, becoming involved in their commercial exploitation.[99] In the second decade of the century he returned to

---

96 *The Columbian Centinel and Massachusetts Federalist*, 5 September 1801.

97 *Supplement to the Encyclopaedia or a Dictionary of Arts, Sciences and Miscellaneous Literature*, Thomas Dobson, Philadelphia, 1803, 3 vols.: v. 1, p. vi and v. 3, appendix, pp. ii–iv.

98 Marion Vernon Brewington, *The Peabody Museum Collection of Navigating Instruments, with Notes on Their Makers*, Peabody Museum, Salem, 1963, p. 84; also *American Neptune, Pictorial Supplement XVII, Instruments of Navigation*, Peabody Museum, Salem, 1975, plate XXIII. A similar instrument, numbered "11" was in private hands in the United States in 1983; see *Boston Globe*, 10 December 1983.

99 English patent 2945 (1806), "Machine for roasting meat by the power of steam", taken out with the Birmingham scale-makers Bourne & Chambers; English patents 3002 (1807), "Machine for weighing to the amount of ten tons and upwards, in place of the common steelyard, or beams and weight"; 3045 (1807) "Machine for mangling linen and other articles"; 3133 (1808) "Construction of a machine for washing or cleansing linen and other articles"; "Gould & Co" advertised their "Patent Portable Family Mangle" in *Morning Chronicle* (London), 24 February 1809, however, *London Gazette*, no. 16398, 21 August 1810 records the dissolution on 28 July 1810 of the partnership between Chester Gould, John

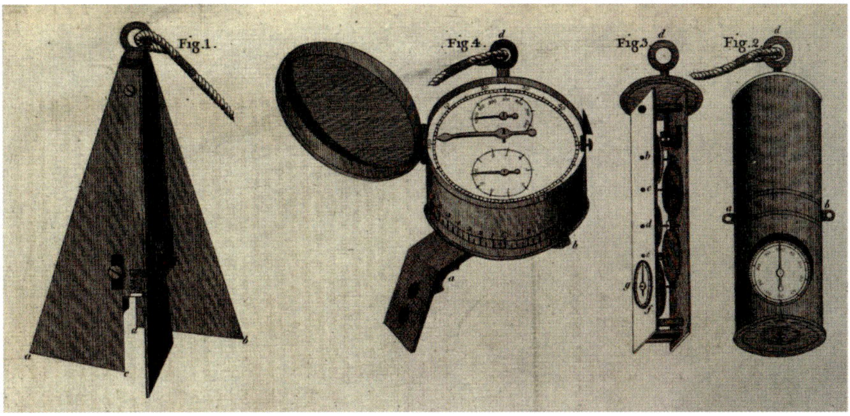

FIGURE 6.6 The second form of the Gould log, from the *Supplement to the Encyclopaedia or a Dictionary of Arts, Sciences and Miscellaneous Literature*, Philadelphia, 1803, vol. 3, appendix, plate 1
COURTESY OF THE NEW YORK PUBLIC LIBRARY

navigational matters taking an interest in Burt's buoy and knipper, a simple and cheap means of sounding.[100]

6 The Massey Patents of 1802 and Later

The one undoubtedly successful design was the log first patented in 1802 by the watchmaker, Edward Massey junior of Hanley, in the Staffordshire potteries

Rusby, and John Wort, of Skinner-Street, Snow-Hill, trading as Gould & Co. Mangle and Press Manufacturers; the other partners briefly continued the business.

100 For Burt's buoy and knipper see D. J. Bryden, "Georgian Instrument Patents – Some Ghosts and Spectres", *Bulletin of the Scientific Instrument Society*, no. 112 (2012), pp. 4–23: 9–10. Gould's association with Burt dates from at least 1815, when Massey complained that Burt, requested to supply a Massey log to the Admiralty Yacht, had "substituted one of MR. Gould's for it"; CUL: RGO 14–31, fol. 137. In 1818 when Massey set up public trials of his sounding machine against that of Burt in London and Liverpool, he consistently referred to it as "Gould & Burt's Buoy and Knipper"; see *Morning Post*, 14 July 1818; also *Public Ledger and Daily Advertiser*, 18 July 1818 and *Liverpool Mercury*, 11 September 1818 and 2 October 1818; Edward Massey, "Comparative Trials of the respective merits of Massey's Patent Sounding Machine and one known by the name of Gould and Burt's Buoy and Knipper", *Philosophical Magazine* 52 (1818), pp. 213–214; and *Journal of Science and the Arts* 6 (1819), p. 135. The title page of Chester Gould, *An extensive table containing eighty thousand azimuths*, J. Ingram, London, 1820, describes the author as "Inventer [sic] of the Sounding Buoy and Knipper, Sold by Peter Burt".

FIGURE 6.7 The Massey log, 1802 patent design
NATIONAL MARITIME MUSEUM, GREENWICH, NAV0728.
PHOTO © NATIONAL MARITIME MUSEUM, GREENWICH

(*Fig. 6.7*). It is a measure of the success of this design that Massey himself protected design developments with further patents in 1834, 1836, 1844, and 1848.[101] Whilst this design was "successful", that success did not come easily or quickly. Examination of the hurdles that Massey had to overcome provide

---

101 English patents: 2601 (1802), "Instrument for taking soundings at sea, measuring reel, mechanical log and current meter"; the specification published in *Repertory of Arts and Manufactures*, 2nd series, 3 (1803), pp. 171–182; "Description d'une machine à sonder les profondeurs à mesurer la marche d'un vaisseau à reconnoitre les sous-courans", *Annales des Arts et Manufactures* 19 (1805), pp. 225–240 closely follows the complete specification; abstracts in *Monthly Magazine or British Register* 16 (1803), p. 154; and *The Naval Chronicle* 10 (1803), pp. 191–192 describe only the sounding machine. 6731 (1834), "Apparatus for measuring the progress of vessels through the water and for taking soundings". 7113 (1836), "Apparatus for measuring the progress of vessels through the water and for taking soundings". 10,210 (1844), "Apparatus for ascertaining the rate at which vessels are passing through the water; also applicable for ascertaining the rate at which streams are running". 12,071 (1848), "Logs and sounding apparatus". The business, and its successors continued to develop the design throughout and beyond the nineteenth century: see Anthony J. Sharp, *Distance Run: A History of the Patent Ship-Log*, Brassbounders (Publishing), Weymouth, 1999, pp. 48–80.

indications of why other alternatives to finding a ship's way at sea with the log-line and half minute glass, fell by the wayside.

As to sea-board trials, Massey was early off the mark. A co-partner in the enterprise, the Liverpool printer and chandler Egerton Smith published a pamphlet on the log and sounding machine in the summer of 1805, filling the final gathering with letters of commendation, indicating that the log and the sounder (which were mechanically and operationally similar) had been trialled by the commander of the gun-brig HMS *Vixen* as early as the summer of 1802.[102] It is unclear whether the second hand report of the approval of the commander and officers of the frigate HMS *Galatea* reflected direct use of the log in 1804, or whether they had observed its use on Captain Killey's Liverpool merchantman, one of the West Indian convoy *Galatea* was supervising. The snow *Lord Nelson* ran a Massey log all the way from Liverpool to Newfoundland in 1804. The Liverpool Committee of the West Indies Association added their weight, as did the local underwriters. When Robert Bill, Massey's patron, financial backer, and co-partner produced a 40-plus page pamphlet the following year, he pointed readers to Smith's tract for a technical description of the design, filling his pages with the reports published by Smith and adding many additional ones.[103] Bill admitted that most of the commendatory letters were from the merchant marine. It was Smith who appears to have pressed local merchant captains to trial the two instruments, and who took the opportunity of HMS *Princess* being in the Mersey estuary to have the sounding machine examined and tested. Some of the merchant captains who had the Massey logs on board, were sufficiently impressed to show them to such naval vessels as they met with. In Naples, Pulteney Malcolm, Captain of HMS *Renown* bought at cost price the Massey log from Joseph Whittle, master of the Liverpool snow *Lord Nelson*, and was so impressed that he kept it with him when he was transferred to HMS *Donegal*. From Portsmouth Malcolm wrote to Bill in London: "I used it during my passage to and from the West Indies, and found it gave the ship's run perfectly correct; a few months since it was broke to pieces towing, I apprehend by a large fish taking hold of it. I wish to procure another before I go to Sea. Do you make them in London? If so I will call at your house".[104] A separate *Second Appendix* was printed later in 1806, for Captain Malcolm

102 Egerton Smith, *Observations on the principle and use of the new patent Sounding Sea Log and Sounding Machine invented by Edward Massey*, E. & W. Smith, Liverpool, 1805.

103 Robert Bill, *A short account of Massey's Patent Log and Sounding Machine, with the opinions of several Captains in the Navy, Merchants Service and Pilots who have made practical use of experimental trials with them*, Blacks & Parry, London, 1806. The pamphlet earned a short and positive review in *Naval Chronicle* 16 (1806), pp. 333–334.

104 Bill, *op. cit.* (n. 103), appendix p. iv.

had introduced Massey to Admiral George Montague at Portsmouth and the latter ordered immediate trials of both log and sounder by several of the vessels under his command, with Massey receiving suitable letters of commendation.[105] A second printing of this appendix was necessary in 1808, with further commendatory letters added as further naval trials had been undertaken by the channel fleet. Significantly the title page carried the notice:

> The Honourable Navy Board have adopted the SOUNDING MACHINE for the use of his Majesty's Navy, and have favoured the Inventor with an order for *Five Hundred* Machines, and a second order for *Two Hundred and Fifty*.[106]

The Royal Navy was fighting a war at sea which involved close patrolling of hostile coast lines in poorly chartered waters, so as to blockade both merchant and naval vessels of the French and their allies. Massey's sounding machine had the great advantage that it would give reliable readings without the necessity of the ship being hove-to. Hence its value to the Navy. The log, in contrast, was primarily of use in providing a more reliable measure of distance run in voyages from port to port. So the positive judgements of the log by naval personnel, however senior, did not result in the Board of Admiralty investing in the design.[107] In the event, Massey was to complain to the Board of Longitude in 1811, that the Navy Board had forced down the unit price of the sounder – paying £3 13s 6d each for the first order and £4 1s for the second, for which he

105 *Second Appendix to the Account of Massey's Patent Log, and Sounding Machine*, Savage & Easingwood, London, [1806].

106 *Second Appendix to the Account of Massey's Patent Log, and Sounding Machine*, William Savage, London, 1808. The additional material, with continuous pagination, forms a "Third Appendix". The copy in the papers of the Board of Longitude has added in mss below that title page paragraph: "Third Order 1000" – referring to the improved sounder with the air chamber re-enforced to prevent collapse at greater depths, and dates to May 1811; see also Robert Clifford, "Extract from a Report made to the Royal Institution on Massey's new sounding machine", *Philosophical Magazine* 40 (1812), pp. 91–99, 97. The 1808 printing, mined by Clifford, was also the source of "Description and use of a sea-log and sounding machine invented by Mr Edward Massey of Hanley in Staffordshire", *Journal of Natural Philosophy, Chemistry and the Arts* 21 (1808), pp. 245–256.

107 Despite the enthusiastic views of some naval captains, the immediate Admiralty view was that the Massey log "did not appear calculated for HM Service": TNA: ADM 12/103 3 February 1803: cited in Robert Gardiner, *Frigates of the Napoleonic Wars*, Chatham, London, 2000, pp. 138–9.

was required to supply a spare set of rotors. Massey's published list price was then £5 5s, so that in essence he was the loser.[108]

Robert Bill's name is no longer associated with the 1808 printing of the pamphlet. An inventive entrepreneur with private means, like Massey he had been born and brought up in Staffordshire. He was introduced to Massey by Egerton Smith. With Massey, the three jointly held the patent rights.[109] Their drive and money undoubtedly helped bring the two instruments to wider attention. However, like many venture capitalists, Bill was distracted by newer interests. He moved to London in 1806 to invest in and become closely involved in a gas lighting scheme; then he erected a factory to manufacture pianofortes made to the 1808 Hawkes patent design. His obituarist notes that he soon parted company with the gas lighting company and that from his investments in the Massey and Hawkes patents "he was the loser of large sums of money".[110] Massey was also developing other ideas. His partnership with Thomas Stanier, to manufacturer his 1808 patent tap, was dissolved in 1813, and the assets liquidated by Stanier.[111] Massey himself was declared bankrupt early 1821, with his assignees taking speedy action to investigate moneys paid out to creditors in Coventry, the potential sale of his watchmaking patent rights, and the improper sale of watches and movements that had been deposited as security for loans. That he was equally speedily issued with a Certificate of Conformity and permitted to continue trading, implies that he and his assignees were able

---

108 CUL: RGO.14–31, fol. 117r, in a letter of 6 March 1811. Nevertheless, advertising his sounding machine over five years later, Massey claimed "The demand for this most excellent Nautical machine has been so great, that the Proprietor has been enabled to reduce the price to THREE GUINEAS AND A HALF" [£3 13s 6d]: *Liverpool Mercury*, 20 September 1816.

109 Massey subsequently indicated to the Board of Longitude, 6 March 1811, that he had parted with 75% of his patent rights to raise the development funds: CUL: RGO.14–31, fol. 117r.

110 *Gentleman's Magazine* 97 (February 1827), pp. 466–8. Bill purchased Andrew Vivian's share in Richard Trevithick's steam carriage, English patent 2599 (1802): see A. W. Skempton *et al.* (eds.), *A Biographical Dictionary of Civil Engineers in Great Britain and Ireland*, Thomas Telford, London, 2002, p. 723. His own English patents were 3418 (1811) "Machine for washing clothes"; 4461 (1820) and 4644 (1822), "constructing beams, masts, yards, &c as metallic tubes"; 4624 (1821), "constructing boats and barges". On Bill and the William Hawkes piano patent, 3154 (1808), see *Monthly Magazine and British Register* 29 (1810), p. 116. The partnership with Egerton Smith was formally dissolved at the end of December 1806: *London Gazette* 16158, 28 June 1808.

111 English patent 3168 (1808), "cock for drawing of liquors"; *London Gazette* 16722, 20 April 1813; *Staffordshire Advertiser*, 17 July 1813.

to come to a financial reconciliation; though they still sold his stock-in-trade of watches.[112]

One hurdle that Massey did not have to cross, was access to manufacturing skills. He had trained as a watchmaker under his father, and he was based in an area with ready access to a pool of skilled mechanicians and all the necessary tools and materials. He was not, however, close to the sea, so it is unclear what stimulated him to design either a log or a sounding machine, both using similar forms of gearing. So far as the log is concerned, there is anecdotal evidence that he tested out his logs on the canal at Newcastle-under-Lyme, to which he had moved in 1806. He later moved production of both his horological and nautical instruments to Coventry, and only later still to the Clerkenwell district of London, both locations where the compartmentalised watch and clock trade businesses provided a pool of highly skilled labour.[113]

Whilst Robert Bill may have quickly tired of supporting Massey's endeavours, Egerton Smith did not. Indeed, when he expanded his printing business into newspaper production, he used his own advertising columns to publicise the activities of his chandlery and instrument-making business, pushing the claims both of his own patent compass, and of the Massey sounder and log, in which he had a substantial financial interest.[114] When Massey himself advertised his new patent marine chronometers in 1815, he included copy relating to the sounding machine and the log, but whereas Smith had described both as "approved and adopted by the Navy", Massey more honestly indicated the naval contracts and awards from the Board of Longitude for the sounder, but of the log he was more circumspect:

> has also received the united testimonials of many of the most distinguished Naval Officers, as an instrument of unrivalled accuracy; and affording the means of keeping a Ship's Dead Reckoning more correctly than by any other known mode.[115]

---

112 *London Gazette* 6 February, 24 March and 24 April 1821. Massey's stock in trade consisting of "gold and silver watches, chains, seals, keys, &c." was auctioned: *The Times*, 9 July 1821; also *Morning Chronicle*, 12 July 1821; the assets of the nautical instrument making business, appears to have been either totally separate, or hidden!

113 Alan Treherne, *The Massey Family; Clock, Watch, Chronometer and Nautical Instrument Makers*, The Borough Museum Newcastle-under-Lyme, Newcastle-under-Lyme, 1977, p. 8.

114 For example: *Liverpool Mercury*, 12 July 1811, 29 May 1812, and 19 February 1813, where the Massey patent log and sounding machine are given a degree of typographical prominence. In later advertising, the log and sounder were merely listed among the many products sold: e.g. *ibid*, 12 December 1817.

115 *Liverpool Mercury*, 19 and 26 May 1815 and 9 June 1815. The Massey & Windham partnership, subsequently advertised the patent sounder and log as "approved and adopted

In time, Massey did supply logs to the Admiralty, but even when the first patent had run its term, they do not appear to have been issued as a matter of routine.[116] Individual naval commanders convinced of their efficacy, made personal purchases.[117] Yet the log-line and glass method of recording speed and thus distance travelled remained standard naval and merchant practice. Why?

7 Failure of Improved Logs to Displace the Log-Line and Glass

Navigational practitioners could readily dismiss as fantasy the drawing-board designs of enthusiastic landsmen. The Board of Longitude, whose academic members had a preference for navigational solutions based on astronomical observation and computation, was generally antipathetic to mere instrumental propositions. Furthermore, however good a design might appear in principle, it had to be physically realised, then tried and tested in a relatively hostile environment by independent and reliable authorities, before the Board would consider making a judgement. Since a patented design would in principle restrict general availability, the Board could and did refuse to consider them. It was only after Massey's sounder and log had been tried, tested and found satisfactory by the navy, that they were willing to consider making a judgement.[118]

by the Honourable Commissioners of his Majesty's Navy": *Naval and Military Gazette*, 21 December 1833, and as "Universally adopted in the Royal Navy": *Greenwich, Woolwich and Deptford Gazette and West Kent Advertiser*, 30 July 1836.

116 For the 1823 Arctic Expedition, Edward Sabine FRS, sailing on HMS *Griper* "obtained from the Admiralty a supply of the logs invented by Mr. Massey": Edward Sabine, *An account of experiments to determine the figure of the earth, by means of the pendulum vibrating seconds in different latitudes*, John Murray, for the Board of Longitude, London, 1825, pp. 436–437; see also Edward Sabine, "Hydrographical Notices", *Philosophical Magazine and Journal* 67 (1826), pp. 332–339, 421–439: 421–2; also Douglas Charles Clavering, "Journal of a Voyage to Spitsbergen and the east coast of Greenland in His Majesty's Ship Griper", *Edinburgh New Philosophical Journal* 9 (1830), pp. 1–14: 2–3.

117 Evidence of Captain H. T. Austin RN in *Report on Steam Vessel Accidents*, Board of Trade, London, 1839, appendix pp. 157–158. See also the praise of the Massey log in "Naval Improvements of the nineteenth century, second notice", *United Service Magazine* (1842) part 3, pp. 492–507: 503–504.

118 CUL: RGO 14–07, Minutes of the Board of Longitude, 5 March 1807: "as he has taken out a patent for those machines they cannot reward him". But at end of the minutes, and in consequence of several certificates from naval officers "of great respectability ... it may be proper to consider the subject". The Board had earlier refused to consider the design for a sextant with built-in artificial horizon proposed by John Syeds – English patent 1789 (1791), and an artificial horizon proposed by Henry Ould, English patent 1842 (1791): CUL: RGO 14/6 Minutes of 3 March 1791: "the Board having understood that both these persons had taken out Patents for their inventions, and thereby secured to themselves the

As Massey's case indicates, bringing a plausible design into production at any scale, required a significant investment in time and money. Purshall (1714) had in principle, a workable (but even then hardly original) idea of a wheel fitted through the keel of the ship itself, towards the stern, and used to measure distance "as exactly as a Perambulator will do the Roads by Land .... From the Wheel there must be some Clock-work with an Index that shall shew every Mile or League that the Ship has run .... Any Body that can make a Perambulator will make this." In 1720 Purshall sought support to have such a machine made, tried, and proved on a ship, offering a tenth of any reward that the Longitude Commissioners might make.[119] What is clear from the trials of other designs that measured distance through the turning of an impeller, is that designers became aware that the number, exact shape and angle of the blades, was a key to operation, and that both theory and the results of trials were needed to calibrate the gearing necessary to record the equivalent distance travelled. In the end, as with the conventional log, distance run still included a degree of fudge, to take into account the very same factors that were known to influence distance run – such as currents, a following sea, yawing as the wind-shifted.

The mechanical "perpetual log" was intended to run twenty-four seven. There were two general forms, those that were integral, and needed to be hauled in and read when a ship changed tack, or whenever the navigating officer wanted a reading of distance run; and those which were intended to provide a ship-board reading of distance run and instantaneous speed. Foxon, whose impeller, like that of Russell, was spiral in form, obtained a ship-board reading by transmitting the rotation of the impeller along a twisting rope to an inboard spindle working the count mechanism. The system worked adequately in smooth water and moderate weather, but there were problems, for example "when the Worm or Log is driven into whirlpools caused by the wake of the Ship it stops, which slackens the Log-Line & the Machine is for that time at a stand, & then the Ship at last going forward the Stray-Line is stretched again & sets the Machine a going."[120] Gower, who may have seen a Foxon log in operation on an East India Company vessel, was of the opinion that "the mode of having a line going from its [the impeller] axis to wheel work on board the ship being very erroneous and inconvenient for measuring distances".[121] His own improved patent log (1807) is an example of a perpetual log with integrated

advantages arising from them, did not think themselves at liberty to order any reward for their inventions".

119 *London Gazette* 5906, 22–26 November 1720. Also see above n. 29.

120 Hoare, *op. cit.* (n. 81), p. 751.

121 Gower, *op. cit.* (n. 22), p. 13.

recording system, like the Savery design known to Newton. However, like the Massey log, such designs had to be hauled in-board to take a reading. Saumarez reported criticism of the Savery instrument:

> Halling on Board to view the Index, and reputting it off again from the side of the Ship into the Sea with great Care and Precaution, least it should be Bruised, which proved Cumbersome and Troublesome to the Sea Men, and of no Dependance to be made upon, much less when the Ship did Veer and Tack about, or kept to the Windward, and that it was useless in the Storm or rough Sea.[122]

The two London instrument makers invited to advise the Society of Arts on the Leguin design of 1787 concluded "from the minuteness of the several parts of the machine it will be liable to be soon deranged by anything that it may meet with in the water".[123] Gower's 1807 design included a davit used to keep the log clear of the disturbance of the wake, and to protect it from being damaged when hauled in-board.[124]

In war time, whether chased or being chased, a naval captain wanting to judge the effect of a change in sail plan, would need to know his speed at any time, for which the integrated log was not designed. In 1800, Gould changed the integrated log of his first patent (2405) to that of a twisting rope working an in-board recorder in the second (2458), a step that was approved of by two American merchant captains, who were particularly enamoured of a perpetual log that they could read from the comfort of the cabin.[125] Massey's log being of the integrated form, may well have mitigated against immediate British naval acceptance.

The Guerimand patent of 1776 is for an in-board read-out. The specification is in the form of an annotated drawing, but without scale or any indication of dimensions. The detail is quite insufficient to allow replication. A horizontal paddle wheel with a half-housing is fixed beneath the vessel, and via a series of "lanthorn pinions" the motions of the axle, driven by the vanes as the vessel passes through the water, is communicated to deck level, where gears record the distance travelled. The vertical shaft transmitting the motion of the paddle wheel is surrounded by a moveable frame, with what might be interpreted

122 De Saumarez, *op. cit.* (n. 30), p. 10.

123 Bryden, *op. cit.* (n. 21), p. 174.

124 Gower, *op. cit.* (n. 24) p. 130 and fig. 5 plate 4.

125 *American Citizen and General Advertiser* (New York), 1 June 1801; reprinted in Gottesman, *op. cit.* (n. 87), pp. 245–246. See also *Directions for using Gould's Nautical Perambulator*, *op. cit.* (n. 94).

as roller-bearings, placed between it and an outer fixed frame. The patentee exhibited a model of the design to the Board of Longitude in the summer of 1781. The response of the Board was non-committal. Guerimand was to show the design to "experienced sea officers" and only return to the Board when he was able to produce certificates indicating the results of sea trials.[126] Gottlieb's 1791 design had the operative impeller "fixed to the side of the keel, and communicates through the bottom of the vessel by a long copper tube, and shews the ship's rate .... on a dial plate fixed to any convenient part on board". The design was fitted to *Carteret* and *Westmoreland*, the first two of an intended order of a dozen newly designed, lightly armed, swift boats commissioned by the government for the Post Office to carry mail between England, North America and the West Indies. I presume that they were not judged a success, for Gottlieb's later advertising deleted reference to his fitting the design to the Post Office packet boats.[127] One of the problems with all forms of perpetual log, was the impeller becoming clogged with weed. One way of dealing with this was included in the Baines 1816 patent, which required a form of a grooved track fastened to the external skin of the vessel leading to a point near the stern post where the operative part of the perpetual log was suspended. The track allowed the whole mechanism to be readily lifted on deck by a flat chain for inspection, maintenance and the clearance of any attached weed. The axle that the impeller turns, is cut with an endless screw that operates a small gear train. That motion is transmitted via a universal joint, using that track with a rod and a further series of universal joints linked by chains, to a read out mechanism on deck. The fitting of the track during the 1818 conversion of the Hull-built whaler to become the Ross expedition HMS *Isabella*, earned criticism expressed in a pseudonymous letter to *The Times* penned by "a Practical Seaman" who signed himself "Navigator":

> it is no new invention, for the Americans have made frequent trials of such machinery, also the dial plate on the top, and found it to err .... Such a water-wheel would answer provided the ship moved on a perfect plain; that there was no heave of the sea, or swell; and that the vessel never gathered stern-way, or went backwards; also provided there was no current. ... there is nothing equal to the common ship's log for ascertaining a vessel's run; the line well measured, the glass correct, and experience to make the necessary allowance.[128]

---

126 CUL: RGO 14/5, *Board of Longitude Minutes*, 21 July 1781.

127 Bryden, *op. cit.* (n. 17).

128 *The Times*, 27 March 1818. The relevant paragraph was reprinted in *Hull Packet*, 31 March 1818. Andrew Mackay, *The Complete Navigator*, 2nd edn., Longman, Hurst, Rees

The innate conservatism expressed by "a Practical Seaman" provides one explanation for the slow acceptance of alternatives to the log-line and glass method of measuring a ship's way. That reluctance to embrace change was expressed by Saumarez in 1725:

> I am well aware, *that he who decries an old Custom, seldom meets with Success, even though what he advances against it may be very reasonable.* This seems to be the Case between the *Marine Surveyor* and the *Log*; for the latter having been long in use amongst the Seamen, it may be with Difficulty the former will be received.[129]

and by Smeaton in 1754 explaining that he had been unable to make further sea trials:

> partly from the indifference, with which I found such a contrivance as this, even if brought to perfection, was likely to be received by seamen; who, in general, do not seem to be over-fond of making trial of new instruments, especially if proposed by landsmen, as, in derision, they are pleased to call us.[130]

The sea can be a very hostile environment. The navigator has to sail his vessel day and night in all circumstances from dead calm to working against adverse winds, coping with inclement weather and conflicting currents, to battling to survive in storm force winds. Whatever was going to replace the log-line and

---

& Orme *et al.*, London, 1810, pp. 11–13 outlines recent designs for perpetual logs, including that of the Philadelphian Francis Hopkinson, which was well known in Britain: see Francis Hopkinson, "Description of a machine for measuring a ship's way through the sea", *Transactions of the American Philosophical Society* 2 (1786), pp. 159–166. The "American" might have been patentee Chester Gould: see above (n. 86); or the unnamed "ingenious American, now in this city, [i.e., London, who] has, we are assured, invented a perpetual log, and the simplicity of the construction promises to be of great utility. It is to be attached to the rudder, so as to shew upon the deck the progress of the ship", reported in *Stamford Mercury*, 12 April 1799 from London 11 April 1799; and in other provincial papers. There were three relevant American patents: Amos Whitmore, "Perambulator for measuring a ship's way", 19 September 1796; Bennet Lies, "Sea Pedrometer", 20 June 1815; John S. Donlevy, "Lexodromic Pulley, for measuring ship's way", 21 September 1815, though no awareness of them has been located in British publications. Inventive patentee Timothy Kirk of Yorktown, Pennsylvania, exhibited a prototype perpetual log in Baltimore in 1808, but it presumably failed to meet expectations: *The Clarion* (Nashville, TN), 2 February 1808; see also *Vermont Centinel* (Burlington, VA), 27 January 1808.

129 Saumarez, *op. cit.* (n. 33), p. 56.

130 John Smeaton, "An account of some experiments upon a machine for measuring the Way of a Ship at Sea", *Philosophical Transactions of the Royal Society*. 48 (1755), pp. 532–546: 544.

glass, whose inadequacies were widely appreciated, had to be at least as accurate and at least as reliable as a method, that with all its failings, had stood the test of time. Supervised trials at sea might well demonstrate the potential of a new method, but as Phipps stated in 1773 when he expressed a preference to the Bouguer log design, "it is not to be determined that the observations of a single voyage can be sufficient to determine the merit of any instrument, particularly one of so much consequence as the log." Phipps's closing comment on the Foxon log is equally important in reflecting the realities of ship-board instrumentation:

> But notwithstanding the advantages which will make it very useful and worth having, I doubt much whether it might ever be substituted entirely in the room of the common log. Machines easily repaired or replaced have advantages at sea, which should not lightly be given up for others more specious.[131]

Simplicity was one of the factors noted by the contributor to the log entry in *Encyclopaedia Metropolitana:*

> Log, a nautical term, implying a machine of some kind for ascertaining the velocity or rate of sailing of a ship .... A variety of contrivances have been made to obtain this object .... Complexity has in many cases, destroyed the laudable views of the projectors. Every apparatus used on ship-board should be simple; for the sailor is placed in too many circumstances of difficulty and danger, to admit of his attention to minute and complicated pieces of mechanism; added to which, every contrivance of such kind contains within it the infallible seeds of embarrassment and decay, from the peculiarly uncertain conditions to which it is exposed .... A variety of ingenious contrivances have also been made to obtain the desired velocity by means of wheel work, but all are open to the objection before adverted to, of being too complicated for such an uncertain and troublesome life as the sea.[132]

The proprietor of a London nautical academy, writing in 1791 aptly commented:

> The Reward held out by Government for the determining of the Longitude has induced many to attempt it, especially in the mechanical Line, and ...

131 Phipps, *op. cit.* (n. 15), p. 98.

132 Sedley *et al., op. cit.* (n. 19), 21 [1832], pp. 521–3.

> many of them have been visionary, whimsical Men, totally unacquainted with the Difficulties attending such a Discovery.[133]

Among those listed in the appendix are a goodly number of such "arm-chair" projectors. It is not mere antiquarianism to include them, for they underline the fact that outside the experienced and informed marine community, there was a wider awareness of the inadequacies of "dead reckoning" with the log-line and sand glass. Despite the totally impractical design, that published anonymously in the *Gentleman's Magazine* in 1755,[134] like the 1813 "Marine Automaton" of Bristol perfumer William Pope,[135] and the embryonic 1835 suggestion of a "perambulator to measure a ship's way" by the animal rights campaigner and serial inventor Lewis Gompertz,[136] with others of a similar ilk, reflect that the need for better instrumental means of measuring the speed of a ship was appreciated by the lay public.

There was every reason for the designs of landsmen to be dismissed by those who spent their lives at sea, battling the elements. The gap between a prototype that might perform tolerably well towed behind a barge on a canal, or within a sheltered estuary, and one that could be routinely handled by the ordinary crewman in extreme conditions, was frequently too wide to be bridged. Simplicity was a key factor in the Admiralty decision of 1817 to prefer the Burt buoy and knipper to Massey's hitherto dominant mechanical sounding machine. Quite independently, commanders of various naval vessels tried the two competing designs, finding that they gave very similar readings. However, each came down in favour of Burt's cheap and simple design. There was little danger of it going out of order, and even if it did, repairs could

133 More, *op. cit.* (n. 4), 9th edn., B. Law & Son, London, 1791, p. 228; though first published in 1772, the comment first appeared in the edition cited.

134 "The description and use of a machine for measuring a ship's way", *Gentleman's Magazine* 25 (1755), pp. 23–24.

135 English patent 3751 (1813), "Instruments for ascertaining a ship's way at sea, and assisting in determining the longitude". On Pope, see Bryden, *op. cit.* (n. 21), pp. 192–193. Augustus de Morgan considered him "one of the crackedest men of his time"; note in his copy of William Pope, *The triumphal chariot of friction*, for the author, London, 1829: University of London, Senate House Library, press mark [DeM] O [Pope] fol. SSR. See also the devastating critique of the book in A. De Morgan, *A Budget of Paradoxes* ed. S. De Morgan, Longmans, Green & Co, London, 1872, pp. 166–167.

136 Lewis Gompertz, *Index to 38 Inventions*, for the author, London [1839], p. 3, also Lewis Gompertz, *Mechanical Inventions and suggestions on land and water locomotion*, W. Horsell, London, 1851, pp. 64–65. Gompertz's design suggestion may be dated to 1835 from mentions in the press; see *The Reformer, and Herts. Beds. Bucks. Essex, Cambridge and Middlesex Advertiser* (Hertford), 3 November 1835, also *Suffolk Chronicle*, 14 November 1835.

be readily undertaken on board, whilst the Massey design was mechanically complicated.[137]

Naval warfare during a significant part of the Hanoverian era was an added complication. A recent study focused on Massey's sounding engine, argues that the seaman using it would not be expected to have sufficient numeracy to read the dials, the device would have to be taken to an officer to examine in the light.[138] Whether or not the numeracy of naval crewmen was so low as to preclude the reading of figures on dials, requires closer investigation, but what has been argued for the slow acceptance of the sounding engine is equally applicable to the mechanical log. What may be more relevant is an appreciation of the naval activity in which the Massey sounding machine had to prove its effectiveness. The smaller and handier vessels of any blockading offshore fleet used the cover of darkness to undertake inshore patrols in poorly charted foreign waters, seeking information about ship movements, and reporting opportunities to cut-out enemy vessels or send in landing parties to destroy naval or military posts, and commercial cargoes. All this required that the vessel showed no lights at all. Measuring depth with the conventional sounding lead or speed with the conventional log, only required touch. This is the context well understood at the time by several designers of compasses and compass binnacles, where a feature was shielding of the night illumination of the compass card, to minimise, if not totally reduce light spillage that might alert the enemy of the presence of an interloper.[139]

---

137 *Copies of Reports of Experiments made by order of the Lords Commissioners of the Admiralty at the request of the Board of Longitude, for the purpose of ascertaining the superiority of Burt's Sounding Buoy and Knipper over Massey's Sounding Machine. With letters from Captains and Masters of Men of War, and Captains in the service of the East India Company, and other Merchant Ships*, for the inventor [Peter Burt], London, 1819, pp. 3–10.

138 James Poskett, "Sounding in Silence: Men, Machines and the Changing Environment of Naval Discipline, 1796–1815", *British Journal for the History of Science* 48 (2015), pp. 213–232.

139 For example, English patent 2581 (1802) which included a transparent or luminous compass: see "Observations of the Patentees" addended to the published specification in *Repertory of Arts and Manufactures* 2nd series 1 (1802), pp. 330–332: 332 notes that their design emits "less light on the sails in chase, or when pursued". Grant Preston's 1812 binnacle and lamp: see Bryden, *op. cit.* (n. 21), pp. 197–198. See also English patent 3540 (1812), "Lighting the binnacle compass"; this advantage not stated in either the specification or the patentee's observations: see *Repertory of Arts, Manufactures and Agriculture*, 2nd series, 21 (1812), pp. 201–202, but is noted in the illustrated advertising leaflet: *The description of a Binnacle Compass illuminated by prismatic reflection, by his Majesty's Royal Letters Patent*, [for P. & G. Dollond], London, 1812, pp. 2–3: "the Apparatus being enclosed, the light is prevented from appearing in the Night to any person except the Steersman … prevents … light from appearing on the deck".

In parallel to the hostile environment of the sea, is operating activity. Heaving the log was a routine task, carried out by seamen, at the command of officers, and reported by word of mouth. The Captain and his officers, or the Master and his Mate(s) on a merchant vessel, looked after and used the delicate octants and sextants, and chronometers, using their measurements to compute latitude and longitude. For a routine task, delegated to ordinary seamen, the tools needed to be able to withstand rough handling, including being regularly and frequently thrown into and pulled out of the sea. Any loss or damage to the conventional log could be readily made good by the ship's carpenter, drawing on existing stores – a vital factor for a naval vessel that might be at sea on blockade duty for months on end, replenished from store ships, permitted to return to a naval dockyard only when in dire need of a refit. It is relevant to note that it was in the years of peace that followed Waterloo, it was largely those naval officers who undertook hydrographic surveying who expressed appreciation of the mechanical log and the mechanical sounder.[140] Their careful measurement activities changed the status of such readings from being routine delegated tasks, to the systematic acquisition of unique data, the quality of which would reflect on their own professional competency.

Generations of midshipmen, in both the navy and the merchant service, had learnt navigation by dead reckoning not just from on-board observation of regular routine use, but through the textbooks that they used to pass the examinations that would give them promotion to full officer status. The numerate mariner cannot but have been aware of the inconsistencies inherent in the varying distance between the marking knots on the long-line, and the fudge implied by using time glasses that ran for less than thirty seconds. One frequently reprinted textbook referred to this practice as "correcting one mistake by another", but accepted that "very often custom prevails over reason".[141] It was not enough for the designer of a mechanical log to solve the technological problems. As Massey was to discover to his cost, entrenched practice formed a formidable barrier to commercial success.

---

140 Edward Belcher, *A Treatise on Nautical Surveying: containing an outline of the duties of the naval surveyor*, P. Richardson, London, 1835, p. 19. In this context note that the Commander of HMS *Hecla* was provided by the Admiralty Hydrographic Office with no less than six Massey patent logs for use on the 1828–1831 West African coastal survey expedition: see Andrew James Webb, "The Expansion of British Naval Hydrographic Administration 1808–1829", PhD dissertation, University of Exeter, 2010, p. 432. See also Thomas Charles Robson, *A Treatise on Marine Surveying*, Longman, Rees & Co., London, 1834, p. 164.

141 Moore, *op. cit.* (n. 4), 3rd edn., B. Law, *et al.*, London, 1776, p. 153.

TABLE 6.1 British log proposals in Hanoverian era, organised by date. All publications printed in London unless otherwise indicated

1714	Conyers PURSHULL, physician, Worcester. PERPETUAL LOG. Submission to Commissioners of Longitude, printed 1715. Brief description of an instrument to measure wind strength, to be used "to find out the exact rate that a ship runs", published 1705.
1717	Henry DE SAUMAREZ, gentleman, Guernsey. PERPETUAL LOG. Book published 1717. Further accounts published by Royal Society 1724/5, with many certificates of performance, also 1730 reporting trial on HMY *William & Mary*.
1722	Joseph GILMORE, teacher of mathematics, Bristol PERPETUAL LOG; THE NAVIVIUM. Book published 1722 – text reprinted 1792.
1731	George REYNOLDSON, upholsterer, York. Patent 534: MACHINE THAT GIVES EXACTLY THE WAY A SHIP MAKES, AND COUNTS HER LEEWAY, tells what speed the winds blow, and proves the different force of the current in any depth. Trials in Thames estuary reported to Royal Society, with certificates of performance.
1735	William BURTSAL MACHINE FOR MEASURING A SHIP'S WAY AT SEA. Design submitted to Royal Society.
1744	William TELFER, watchmaker, Glasgow. MACHINE FOR MEASURING A SHIP'S WAY. Press report of succesful sea trial off Leith.
c.1750	William RUSSELL, administrator in the Levant Company. PERPETUAL LOG. Design reported *en passant* by Smeaton (1754) as "some years since"; had been tried at sea. Developed form tested and discussed by Phipps (1773).
1754	John SMEATON, philosophical instrument maker, London. PERPETUAL LOG published by Royal Society (1755). Naval sea trials in parallel with tests of Gowin Knight's marine compass.
1755	ANON MACHINE FOR MEASURING A SHIP'S WAY described and illustrated in *Gentleman's Magazine*. Theoretical design only.
1767	Peter Perez BURDETT, cartographer, Derby. NEW MARINE SURVEYOR. Press report of sea trials.
1772	William FOXON, carpenter, Deptford. PERPETUAL LOG. Patent 1028: "Machine for measuring a ship's way". Tested and published by Phipps (1773). Examples supplied to and used by Navy and East India Co.

TABLE 6.1 British log proposals in Hanoverian era, organised by date (*cont.*)

1774	James RYMER, RN, surgeon's mate. MACHINE TO NEASURE A SHIP'S WAY. Submitted to Society of Arts.
1776	Jacques GUERIMAND, journeyman watchmaker, London. PERPETUAL LOG. Patent 1133; machine for measuring a ship's way with more accuracy than the log-line, "Marine Perambulator". Model submitted to Board of Longitude 1781.
1777	Johann E. VETTER, lecturer in practical mathematics and military sciences, Prussia. PERPETUAL LOG. Detailed drawing submitted to the Board of Longitude.
1783	Walter BEDFORD, Portsmouth. LONGITUDE CLOCK FOR ASCERTAINING THE SHIP'S VELOCITY. Proposal submitted to Board of Longitude.
1786	George SINCLAIR, Thurso. PERPETUAL LOG. Proposal submitted to Board of Longitude. No design detail supplied.
	Richard Benson WALKER, merchant, Kingsland, Middlesex. MACHINE FOR MEASURING A SHIP'S WAY AT SEA. Certificate of performance submitted to Board of Longitude 1790. Admiralty trials ordered and undertaken – expenses paid.
1787	Estienne LEGUIN, French inventor, trading as optician in London. MECHANICAL LOG, submitted to Society of Arts and subsequently to Board of Longitude. Later re-submitted to French authorities and published (Paris 1803), and from that source (London 1804).
1788	William COOPER, Sheffield. WATER PERAMBULATOR. Outline proposal submitted to Board of Longitude.
1789	Matthew VALLET, French industrial chemist, resident in north west England from 1785. MECHANICAL LOG. Description and drawings submitted to Society of Arts 1796, with reports of seabord trials in Mersey estuary 1790 and 1791. Earlier submission to French authorities.
1790	Felix Antonio CASTRIOTO, London-based, Portuguese journalist. MACHINE FOR MEASURING A SHIP'S WAY. General proposal submitted to Board of Longitude
	J. POINTER, postmaster, Newbury. INSTRUMENT FOR MEASURING DISTANCES AT SEA. Instrument submitted to Society of Arts and Board of Longitude. No design details known.

TABLE 6.1 British log proposals in Hanoverian era, organised by date (*cont.*)

1791	Valentine GOTTLIEB, German immigrant in London from 1776; mechanician and manufacturer. PERPETUAL LOG. Design published (1791) and re-printed (1792). Fitted to two post office packets.
1792	John COOKE, gentleman, Kiltinan Castle, Tipperary, Ireland. PITOT TUBE TYPE. Design proposal submitted to Board of Longitude.
	Richard Hall GOWER, mariner, East India Company, Edmonton. PERPETUAL LOG. Patent 1895: instrument for measuring a ship's way through the water, and for ascertaining the rate of sailing at any time. See also his 1792 book. Sea trials undertaken.
	Benjamin MARSTON, American-born Loyalist, Harvard graduate – 1784 deputy surveyor general of the King's Woods for New Brunswick; destitute in London from 1787. MARINE DRONOMETER TO MEASURE A SHIP'S WAY. Outline proposal submitted to Board of Longitude.
*c.*1795	John BASSETT PERPETUAL LOG. Detailed design proposal found in his papers, published (1830).
1800	Chester GOULD, merchant, Rome, Oneida County, New York State – Emigrated to US but returned to live in England. PERPETUAL LOG. Patent 2405: "Instrument or log for ascertaining a ship's distance at sea." Patented in US 1799 as "Perpetual Log or distance clock". US Navy sea trials reported in US press and in UK and France. English specification printed (1800); abstract in German journal (1801). Patent 2458 "An improved instrument or log for ascertaining a ship's distance at sea". US Navy sea trials reported in US press. Specfication printed (1801); and Philadelphia (1803).
*c.*1800	William Fitwilliam OWEN, lieutenant RN, marine surveyor. IMPROVED FORM OF BOUGUER CONE LOG. Detailed and published (1804).
1801	Thomas HAMILTON, commander of the Sea Fencibles, Margate. PITOT TUBE TYPE. Design published (1801). Positive trial report from HMS *Isis*.
1802	Edward MASSEY, watchmaker, Hanley, Stoke on Trent. PERPETUAL LOG. Patent 2601: "Instrument for taking soundings at sea, measuring reel, mechanical log and current meter". Specification published (1803). Many other publications in newspapers, journals and books. On general sale from at least 1811. Submitted to Board of Longitude in 1806. BoL eventually rewarded Massey for his sounder (improved design

TABLE 6.1 British log proposals in Hanoverian era, organised by date (*cont.*)

	patented 1806). The Navy ordered sounders in bulk; Massey marketed them together. Further patents 1834 and 1836.
1803	J. PIMLOT, Butley Town, Cheshire. PERPETUAL LOG. Design submitted to Board of Longitude. Small scale trials reported.
1804	James SHARPLES, portrait painter, Bath. Patent 2794: "Combinations and arrangements of implements and mechanical powers, and certain principles and forms of tables, useful for surveying and various other purposes" [forms of counters for waywisers and other distance measuring devices, including A SHIP'S LOG and a pedometer for use by horses].
1805	John AUSTIN, retired Glasgow textile manufacturer, now resident in London. SHIP'S LOG – with depth sounder and night telescope. Design submitted to Society of Arts. No details known.
1806	Edward FERGUSSON, retired Lt. Col. E.I.C. (Madras), resident in London from 1804. Method of using his CURRENT LOG (tried on the Thames) submitted to Board of Longitude. No design details provided.
	Hugh HAMILL Junior, merchant, Dublin. PITOT TUBE TYPE, design proposal only – published (1806).
1807	John Whitley BOSWELL, Irish-born serial inventor, and ship designer, London. SPRING-LOADED PERPETUAL LOG published (1807), reprinted (Paris 1808). Theoretical proposition only.
	Richard Hall GOWER, former mariner, East India Company – Naval Architect. PATENT PERPETUAL LOG – up-dated design of 1792 patent, published by Gower (1807) after trials.
	Charles GRANT, Vicomte De Vaux, refugee French monarchist general, resident in London. Patent 3088: "Machine to show the latitude and longitude at sea, also for weighing objects, MEASURING SPACE, OR A SHIP'S COURSE, OR KEEPING ACCOUNT UPON DIALS and cosmographical columns, SHOWING ALSO THE LEEWAY OF A SHIP" – partly applicable to other purposes. METHOD OF ASCERTAINING THE RATE OF A SHIP BY MEANS OF CLOCKWORK submitted to Board of Longitude in 1806 and in revised form, 1807.
1809	James BURNEY FRS, captain RN (unemployed), London. SPRING-LOADED PERPETUAL LOG design published (1809).

TABLE 6.1 British log proposals in Hanoverian era, organised by date (*cont.*)

1813	William MARTIN, natural philosopher, poet and eccentric inventor, Northumberland. SPRING-LOADED LOG. Model and drawing submitted to Society of Arts.
	William POPE, perfumer, Bristol. SPRING-LOADED PERPETUAL LOG. Patent 3751: "Instruments for ascertaining a ship's way at sea, and assisting in determining the longitude". Marine automaton, with current log and compass card.
1814	Henry BROWSE Junior, gentleman, Paignton. PERPETUAL LOG – design proposal, and model submitted to Board of Longitude
1816	Robert Raines BAINES, glue manufacturer, Hull. PERPETUAL LOG OR SEA PERAMBULATOR. Patent 4082. Specification printed (1820); non-technical synopsis (1820) – reprinted (New York 1820). Positive performance reports from naval and merchant vessels. Tried on Ross 1818 NW Passage expedition. Submitted to Board of Longitude 1819. Baines trading in Hull as patent log manufacturer until 1828.
	John TYSON, London. SHIP'S WAY AT SEA FROM PORT TO PORT WITHOUT HEAVING A LOG. Broad proposal submitted to Board of Longitude. Fails to provide design details.
1819	James BAMBER, Ballymoney, Co. Antrim, Ireland. Proposal for a simplified MECHANICAL LOG submitted to Board of Longitude – unclear if prototype was made, rather than merely and inadequately described. Not subjected to trials.
*c.*1820	Arthur Batt BINGHAM, captain RN. PERPETUAL LOG – "MARINODOMETER". Model submitted to Board of Longitude with certificates of performance of trials on HMS *Dover*. Also reported as fitted to the keel of PS *Tourist* (1824).
1822	Arthur HODGE, inventor and tabulator, from St Croix, West Indies, in London since 1805. PERPETUAL LOG, RESEMBLING A FISHING-ROD. Design submitted to Board of Longitude.
1823	Mark COSNAHAN, merchant, ship-owner, and banker, Isle of Man and Liverpool. SPRING-LOADED PERPETUAL LOG. Patent 5130 (1825). Apparatus for ascertaining the way or leeway of ships and other vessels – "marine computist". Abstract of specification published (1826). What the device would achieve, but not how, published 1823.

TABLE 6.1 British log proposals in Hanoverian era, organised by date (*cont.*)

Date	Proposal
1827	James HOOKEY, midshipman, RN. IMPROVED SHIP'S LOG submitted to Society of Arts awarded a silver medal for the design of a fish-shaped log slip. Trials on HMS *Supply*. Published (1827) – re-printed (1828).
	James LAMB, Liverpool. PERPETUAL LOG submitted to Society of Arts. No design details known
1828	Francis HIGGINSON, Mate RN. SPRING-LOADED LOG and LEE-WAY INSTRUMENT – proposal submitted to Board of Longitude, without technical detail. Submitted to Society of Arts, with reports of trials on cutter HMS *Arrow*.
1830	John ELVEY, Millwright, Canterbury. MECHANICAL LOG to tell a ship's way and leeway without heaving the log. Press report that model of the design completed 1831. Positive result of sea trial 1832 reported – further improvements required.
	Thomas GRAVES, lieutenant, RN. CONICAL LOG with flotation compartment. Trialled by designer. Published 1831
*c.*1832	George EVANS, commander, RN. PITOT TYPE LOG fitted to the new steam vessell HMS *Rhadamanthus*, "not found to answer".
1834	Edward MASSEY Senior, watchmaker, King Street, Clerkenwell, London. MECHANICAL LOG. Patent 6731. Apparatus for measuring the progress of vessels through the water and for taking soundings. With revised specification patent 7113 (1836). Update of 1802 patent. Widely advertised and sold.
1838	Leandre CLEMENT, watchmaker, Rochefort, France; engineer, London. Patent 7780 PERPETUAL LOG – Ascertaining and indicating the rate of vessels passing through water. Also patented in France as "MONTRE À SILLAGE", postive reports of naval trials published from 1837 in France. British Naval trials reported from 1842.

CHAPTER 7

The Transmission X-Ray Microscope: Super Microscope for Biology or Yet Another Niche Instrument?

Andreas Junk

The concept of the transmission x-ray microscope (TXM) was introduced to the scientific community by Günter Schmahl and Dietbert Rudolph in 1967. They devised a microscopical setup based on Fresnel zone plates, which could make images of biological specimens in the region of the water window (i.e., the region of the electromagnetic spectrum between 2.5 and 4.5 nm in which water is transparent to x-rays). Such a microscope would have overcome the paradigm of "unfocusable x-rays" and also would have been a super microscope for biology, being able to convey three-dimensional information of thick samples of a few micrometers.

The development allegedly started after a symposium on the use of holographic techniques as devised by Dennis Gabor. Schmahl and Rudolph, then assistants at the University Observatory Göttingen, fabricated the first zone plates in their spare time using laser light.[1] A zone-plate telescope carrying such zone plates was used as a test environment for the production procedure.[2] After successfully making images of the Sun's x-ray spectrum, the fabrication techniques were enhanced to make lenses for a microscope.[3]

Although the setup of the TXM was completed and tested by the mid to late seventies, the results from the images did not convince biologists to use the new instrument. Even after the completion of a series of high-resolution images of malaria-infected red cells, the instrument was not recognised as being necessary or paramount for research in the life sciences. Schmahl and Rudolph did receive a few prizes for the TXM's development, yet they were unable to get the

1 Günter Schmahl and Dietbert Rudolph, "Lichtstarke Zonenplatten als abbildende Systeme für weiche Röntgenstrahlen", *Optik (Jena)* 29 (1969), pp. 577–585.

2 Dietbert Rudolph and Günter Schmahl, "Holographically made zone plates for use in x-ray telescopes", in F. Labuhn and F. Lüst (eds.), *New Techniques in Space Astronomy*, Springer, Dordrecht, 1971, pp. 205–206.

3 Heinrich Bräuninger, Hans Jürgen Einighammer, Johannes Feitzinger, Henner Fink, D. H. Höhn, H. Koops, Gerhard Krämer, U. Mayer, Gottfried Möllenstedt, and Manfred Mozer, "EUV and Soft X-Ray Images of The Sun on March 11th 1971", *Solar Physics* 20 (1971), pp. 81–84.

 | DOI:10.1163/9789004689107_009

life sciences to show considerable interest in the instrument. It seems, biologists rather stuck to their traditional methods of research because the information gained by using the TXM did not seem to be worthwhile. Today the TXM is used at various synchrotron storage facilities in the field of material research.

The case study presented in this paper is based on oral history interviews, which I conducted with the TXM's developer and several of his former colleagues. My aim was to evaluate why an instrument, which could deliver high-resolution images of red blood cells deteriorating because of a malaria infection,[4] was not widely used in the life sciences but – at the time of my investigation – seemed to be stuck for analysis in the material sciences.

After introducing the technology of the transmission x-ray microscope (TXM) and in particular, its new optics, I will present excerpts of the interviews and show the direction they pointed me in when I conducted my research. I will offer a number of categories in which aspects of success and failure can be seen. Although I will talk about a working instrument and the successful implementation of a new kind of diffraction optics, I will conclude that this instrument is still a failure. I see it as a failure mainly because of the promises it made, which could not be kept; the weak influence of its developers, whom were apparently unable to propagate its use outside of their own community and did not produce compelling images that would have suggested broader use; and last but not least, the failure of technicians to create a laboratory version of the instrument.

1 The Setup of the TXM

X-rays are notorious for two properties: They "cannot be concentrated", a fact noted in Röntgen's first publication on the discovery of x-rays;[5] and they penetrate biological specimens, especially tissue, and metals very effectively. Cell biologists have avoided analysing tissue samples with x-rays because the radiation would damage the cells. But from a physicist's point of view, the shorter wavelength of x-ray photons has one obvious advantage – images with enhanced spatial resolution. An instrument such as the TXM, with shorter

4 Cathleen Magowan, Jackson Yeung, John Brown, Ajit Nair, Elissa Meites, Ross Coppel, Narla Mohandas, and Werner Meyer-Ilse, "X-Ray Microscopic Visualization of Specific Labeling of Adhesive Molecule CD36 and Cytoadherence by *Plasmodium Falciparum* Infected Erythrocytes", *Research Communications in Molecular Pathology and Pharmacology* 99, no. 4 (1998), pp. 245–258.

5 Wilhelm Conrad Röntgen, "Über eine neue Art von Strahlen", *Sitzungsberichte der Würzburger Physik.-medic. Gesellschaft* 1 (1895), pp. 1–9.

FIGURE 7.1 A zone plate is a structure of concentric rings, whose diameter increases in a $\sqrt{n}$ ratio relative to the centre area. The width of the outermost zone dr_n defines the limits of spatial resolution

wavelengths than visual light, can reveal more structural details of a specimen. In addition, when radiation in the region of the water window is used, the biological samples can be examined without the use of a contrast agent.

The use of zone plates were Schmahl's and Rudolph's solution to overcoming the stigma of the allegedly unfocusable x-rays (*Fig. 7.1*). Fresnel zone plates had been used long before the TXM as a means to focus visible light by diffraction, after an idea of J. L. Soret and J. W. Strutt,[6] which had been implemented by R. W. Wood to enlarge an image recorded with his personal camera obscura.[7]

6 Jacques Louis Soret, "Ueber die durch Kreisgitter erzeugten Diffractionsphänomene", *Annalen der Physik* 232, no. 9 (1875), pp. 99–113; Robert John Sturtt, *Life of John William Sturtt, 3rd Baron Rayleigh*, Edward Arnold, London, 1924, pp. 88–89.

7 Robert Williams Wood, "Phase-Reversal Zone-Plates, and Diffraction-Telescopes", *London, Edinburgh, and Dublin Philosophical Magazine and Journal of Science* 45 (1898), pp. 511–522.

A zone plate is a setup of a number (n) of concentric rings of decreasing width. The radius (r) of each zone is calculated by the formula

$$r_n = \sqrt{n} \cdot r_1$$

The width of the outermost zone (the "nth" zone) very much defines the zone plate's optical properties. Its width basically equals the lateral resolution. If you wanted to analyse structures in the region of 10 nanometers, for which you would have to use radiation with a wavelength (λ) of 10 nanometers or less, then the outermost zone of the "micro zone plate" should not exceed this particular value. Any objects larger than the wavelength used will be imaged in more detail – e.g., the surface structures of cell tissue (typically in the micrometer size range) can be identified with x-rays (typically having wavelengths in the nanometer range).

The setup of the TXM resembles a traditional visible light microscope as it might be used in a secondary school's biology class. In the normal light microscope, the specimen is usually placed between the condenser and the objective lens; a third lens, the ocular lens is used to form the image in the spectator's eye. The lenses used for visible light are usually made of glass, and the physical effect for image formation is then refraction. The TXM does not use photons in the visible light range (wavelengths between 400 and 800 nm), but x-rays (1 to 100 nm wavelength). Physicists talk about soft x-rays, hard x-rays, and gamma rays in this wavelength region. The refractive index for any sort of glass in the soft x-ray region is below one, so any biconvex glass lens would disperse the light transmitted through it instead of focusing it. Here is where the micro zone plate – a type of Fresnel lens – does the job. It focuses light by diffraction. However, x-rays are invisible to the human eye, so there must be some modification to the eyepiece of the microscope. Instead of using another Fresnel lens as a replacement for the ocular lens, Schmahl and his colleagues initially used x-ray-sensitive photographic film to record the image already formed by the objective lens, the so-called micro zone plate. *Figure 7.2* shows a more elaborate version of the TXM instrument in which the holder for the photographic film has been replaced by a CCD chip in the same spot.[8]

8 Image reprinted from Peter Fischer, Dong-Hyun Kim, Weilun Chao, J. Alexander Liddle, Erik H. Anderson, and David T. Attwood, "Soft X-Ray Microscopy of Nanomagnetism", *Materials Today* 9, nos. 1–2 (2006), pp. 26–33, with permission from Elsevier.

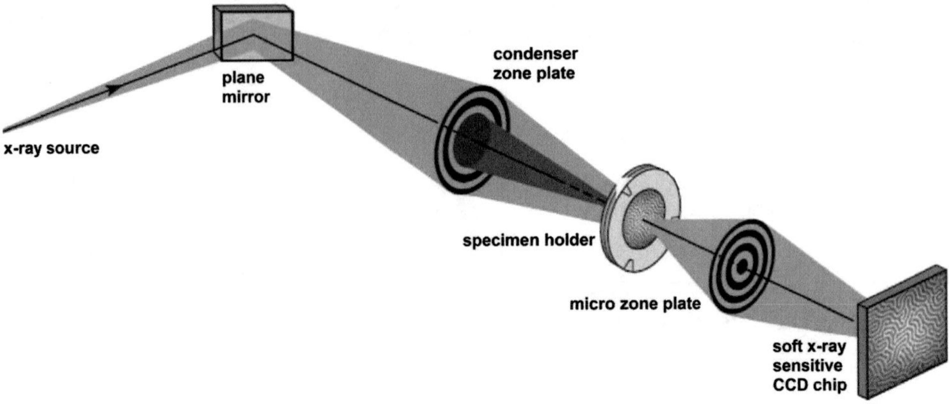

FIGURE 7.2 The modern setup of the TXM. The condenser zone plate (CZP) collimates the x-rays for better illumination of the sample. The micro zone plate (MZP), with more and much finer outermost zones, is used to form the image on the CCD chip. Image adapted from figure 1 in Peter Fischer *et al.*, "Soft X-Ray Microscopy of Nanomagnetism", *Materials Today* 9, nos. 1–2 (2006), pp. 26–33, but altered by the removal of peripheral instruments and the magnet, and by relabelling

2 Astronomy, Oral Presentations, and the Drive to Work the Next Day

My first oral history interview took place with Günter Schmahl, then already emeritus professor and former director of the Institute for X-Ray Physics.[9] My research in the run-up to the interview identified him and his long-time colleague Dietbert Rudolph as the inventors of the TXM. I chose to interview only one of the two, because I wanted my interviews to tell different sides of the story of the TXM's development, and I expected Rudolph's story to be very much aligned with Schmahl's.

Schmahl had finished his PhD thesis in 1966 and was of course not yet in a position to form his own research group. At that time, he was a research assistant at the University Observatory in Göttingen in central Germany, and his main job was to interpret the emission spectra of the Sun and the fine structure of the lithium-6 / lithium-7 lines, because those were considered instrumental to predict the fusion process in the Sun's core. To do so, one had to use grating monochromators but these in turn appeared to be limited in wavelength resolution, mainly due to grating ghosts and the occurrence of stray light. The

9 Günter Schmahl, interview with Andreas Junk, Göttingen, 4 April 2006.

engraved diffraction gratings used at that time could not be improved, so the observatory was always interested to learn about developments in the production of these gratings.

An oral presentation by an external guest in the observatory's colloquium was the moment of nucleation to improve the situation. The focus of the presentation was the use of holographic techniques. Schmahl's recollection of the evening after the colloquium was as follows:

> In the evening after the colloquium, Mr. Rudolph, me, and our wives played bridge until late in the night. In the following morning we didn't have the drive to work as usual and so we discussed the options of using an interference field as a basis to produce a diffraction grating. […] We realised, that we needed a reliable way of information storage and agreed upon the usage of laser radiation combined with photoresists, which were known from semiconductor technologies.[10]

According to Schmahl, he and Rudolph evaluated the holographic fabrication of gratings during the next two weeks, and Schmahl even gave a presentation on the consequences and what would be needed by the observatory to fabricate its own gratings. Above all, they had to solve the problem of illumination, because in a wavelength-selective setup, a lot of light is lost due to diffraction. Schmahl and Rudolph finally agreed upon using critical illumination instead of Koehler illumination, because they would not need an additional diffraction optical element and hence would have more light for the formation of the image. The photoresists to be used were only sensitive to blue or ultraviolet light, and for the holographic approach to work, a laser had to be used. The lasers that were widely available back then only emitted in the red or green region of the visible spectrum, and hence had not ample photonic energy to activate the resists. Schmahl gave an in-house presentation of his idea to use a laser system emitting in the blue region, and he also applied for funding.[11] Funding was declined because the blue laser would be too expensive; he was told to wait until the photoresists were sensitive in the green or red region of the spectrum, which was out of the question for technical reasons.

By a bit of luck and craft, Schmahl's group was able to get their hands on a Spectra Physics laser emitting in the blue region, which was loaned to the observatory by the Heidenhain enterprise in Traunreut. This company was mainly known for precision gears, but also made some efforts in microtechnology

10 Schmahl, *op. cit.* (n. 9). Author's translation from the German.

11 Schmahl, *op. cit.* (n. 9).

with photoresists. Since that branch of the company was an aside to its routine duties, the company basically transferred the research to the observatory. Schmahl and Rudolph were now in a position not only to generate gratings, but also zone-plate lenses. Working on the principle of diffraction, these circular structures could be used as lenses, whereas glass lenses used refraction. Schmahl gave a vivid account of what inspired him to think of this option:

> We knew about one question from an oral exam by Möllenstedt. "What can you use a zone plate for?" and the expected answer "To build an X-Ray Microscope". And that is, what we wanted to do, build those lenses, not only for a telescope, but also for a microscope. That was on one single day.[12]

Of course, one has to be careful reading these accounts; a lot might be a recollection *ex post*. But the general idea to build lenses would probably not be confined to their usage in a zone-plate telescope. The idea that a microscope could be built as well with these lenses would have been a natural thing for a scientist.[13]

Since Schmahl and Rudolph worked at an observatory, they decided to produce their lenses for a solar telescope, which would be mounted on a ballistic missile. The idea to test new optics in telescopes, which usually have longer focal lengths than the lenses in microscopes, came in handy. The fabrication of zone plates was not yet stabilised to a point where a very large number of zones could be recorded with the photoresists. But since the focal length of the zone plates largely depended on the number of zones, a small number of zones would result in a large focal length (and vice versa). The focal length of a zone plate is given by the formula:

$$f = \frac{r_n^2}{n\lambda}$$

with f being the focal length, r_n the radius of the nth and outermost zone, and λ the wavelength used. These missile tests only gave a very restricted amount

12 Schmahl, *op. cit.* (n. 9). Author's translation from the German.

13 As a remark on the usage of these quotations: There is an imminent danger in believing that everything the interviewees say will be a truth. The preparation for the interviews keeps an interviewer into falling into simple traps; the aftermath comprises a close study of other sources to verify or refute what has been said. But this discussion would lead into one about oral history as a method, which I do not want to begin here.

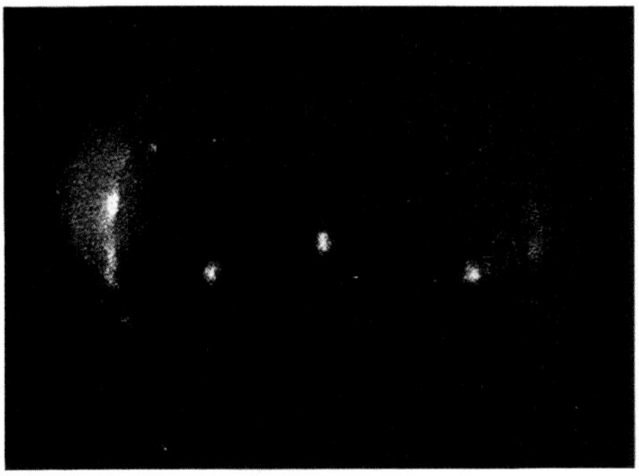

FIGURE 7.3
Experimental proof that zone plates could be used as imaging optics. The image shows the Fe XVII radiation from the Sun, recorded during ESRO Mission CR16 on 11 March 1971
REPRINTED WITH PERMISSION OF THE UNIVERSITY OF TÜBINGEN AND THE ASTRONOMISCHE GESELLSCHAFT, GERMANY

of time for measurements because the time outside the stratosphere was limited to a few minutes. One of the zone-plate telescope prototypes did not even make it into orbit; the carrier missile crashed prior to reaching measurement altitude. In 1972, the concept of a zone-plate telescope was proven successful by taking an image of the Sun's x-ray radiation (see *Fig. 7.3*). The imaging properties were good enough to drive the development of the zone plates' structures into continuously finer outer rings, because the width of these outermost structures were the ones that guaranteed resolution at short wavelengths. The spatial resolution was basically equal to the width of the outermost zone of the zone plates.

3 Using the New Optics for a New Instrument

The successful test of the zone plates in a telescopic setup was the much needed proof that the diffraction optics would work for an imaging instrument, but the actual design of the TXM had yet to be devised. Since the principal setup was no secret, the first zone-plate microscope was put together and tested at the Deutsches Elektronen-Synchrotron (DESY) facility in Hamburg. DESY was able to provide x-rays of variable wavelength with enough intensity to test the optics even though it was not in the perfect wavelength region in which the TXM was aimed to work. Schmahl and his colleagues had to use a lot of time to make their images, because the images could only be shot one by one, and the setup was brought into a vacuum chamber for testing. This meant that after every exposure, the chamber had to be ventilated, the holographic film removed, and then developed. Only after developing the image could the group see whether or not the alignment of the optics had been good enough. After adjusting the optics' holders bit by bit, the film holder was reequipped

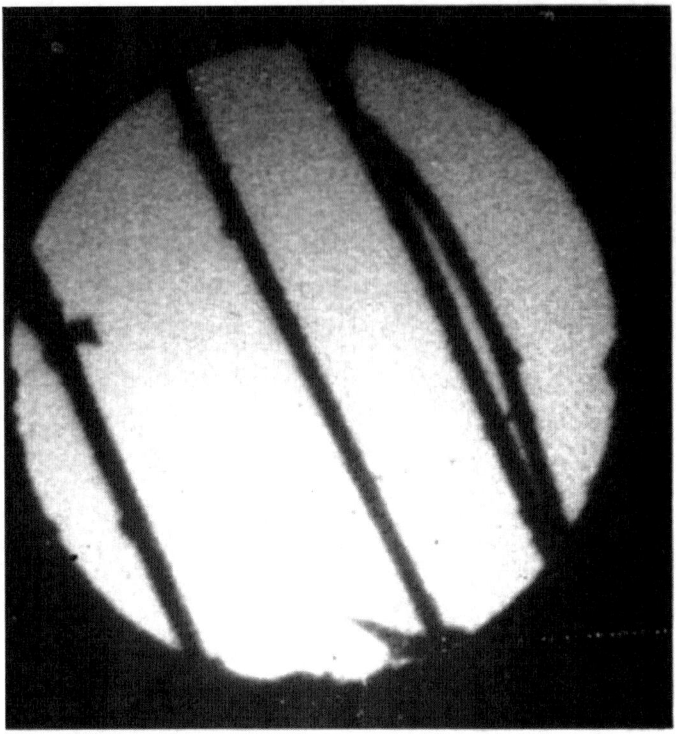

FIGURE 7.4 The first successful recording with the TXM prototype shows a metallic copper grating. The resolution of the details in the image is 2μm. Reprinted from B. Niemann, D. Rudolph, and G. Schmahl, "Soft X-Ray Imaging Zone Plates with Large Zone Numbers for Microscopic and Spectroscopic Applications", *Optics Communications* 12, no. 2 (1974), pp. 160–163
WITH PERMISSION FROM ELSEVIER

with another piece of film, and the chamber eventually evacuated again.[14] That would not have been too bad, if the hard x-rays provided by DESY had not deteriorated the zone plates themselves; the group had not accounted for the tremendous ageing process the optics underwent during this testing procedure. The principal researcher, Bastian Niemann, the group's first PhD candidate in the new field, finally managed to shoot a good image of details of a copper grating at a wavelength of 4.5 nm. The grating fibres were as thick as 10 μm and the spatial resolution was 2 μm (see *Fig. 7.4*). The setup and alignment

14 Bastian Niemann, interview with Andreas Junk, Göttingen, 29 June 2006.

for the optics was tested and proven, but in order to increase the resolution, zone plates with a much larger number of zones had to be produced.

Bastian Niemann, who later became the most seasoned researcher Schmahl had at his disposal in the years to come, was tasked to build a so-called laboratory x-ray source, which would be the basis to expand the field into almost every university laboratory that wanted to join the field. The approaches to generate x-rays were various, and most of them successful in principle. The Göttingen group had modest success with using a nitrogen-based plasma source, from which radiation at 2.47 nm could be gained.[15] (*Fig. 7.4*) The drawback was that these sources could not provide enough photons, so in the end there was no sufficient illumination to make images of any sample. Even if the search for a working laboratory source was a setback, the 1970s brought financial independence when the German Volkswagenstiftung gave massive financial assistance to the group. Schmahl identified a backdoor in the funding criteria and applied for funding for development of a super microscope for biology instead of doing a standardised physical experiment on diffraction optics. Many years after the programme, he learned that the holographic approach was evaluated by a powerful authority: Dennis Gabor himself.

> I learned later on that the foundation got back to Dennis Gabor for consultation, because of our holographic approach. He clearly stated that it need not work out, but if it should work out, the payoff would be really nice.[16]

Having the money was of course a relief to Schmahl's group, because they could advance their production of the optics with largely improved equipment, but at this point the resistance within the observatory became a lot bigger. First of all, it was not obvious anymore that the optics would be produced for the sake of astronomical research, and some thought that the "production" should be left to industry rather than a university.[17]

Schmahl and his group were striving to find a setup that would finally allow only for adjustments. The next series of tests was made from the mid to late seventies on at the ACO storage ring in Paris, which provided soft enough

15 Dietbert Rudolph, Bastian Niemann, Günter Schmahl, and Jürgen Thieme, "Status of a Laboratory X-Ray Microscope", *Springer Series in Optical Sciences* 56, no. 3 (1988), pp. 216–219.

16 Schmahl, *op. cit.* (n. 9). Author's translation from the German.

17 Schmahl, *op. cit.* (n. 9).

x-rays to prolong the zone plates' lifetime and let the group identify the initial troubles in the instrument. But the success here sealed the group's fate at the observatory's premises.

The technical success with the TXM led to the group's receipt in 1985 of the Niedersachsenpreis, a prize offered for scientific developments in the German state of Lower Saxony. Lower Saxony's Prime Minister was given a grand tour of the facilities, and to his dismay, he found the enhanced yet still very much improvised laboratories and offices on the observatory's premises. Shortly after receiving the prize and taking advantage of the new political support, Schmahl and his group moved into a fully renovated building with new equipment and became the "Research Group for X-Ray Optics", which eventually became the "Institute for X-Ray Optics".

The results obtained by Schmahl's group became known to a new community of physicists. The field of x-ray microscopy was stable enough by the beginning of the 1980s to produce conference proceedings with Springer, and Schmahl and his colleagues provided many articles for these volumes. So the dissemination of results, be it about enhanced fabrication techniques, preparation of the samples, or alterations of the setup became known to this community. The work of the group in Göttingen was highly appreciated, and some who trained there, like Werner Meyer-Ilse, went on to run research units in the United States. The field was also recognised by its inclusion in a new section on x-ray optics (written by Günter Schmahl) in the standard teaching volumes for Germany's physics students.

4 Malaria-Infected Red Blood Cells – Iconic Breakthrough or the Ultimate Reason for Failure?

Developing microscopic resolution at a sub-cellular level could have been one of the trademark achievements of the TXM. Spatial resolution at the submicrometer down to the nanometer level was already possible in principle when one used the electron microscope (EM). But the electron microscope was unable to scan biological specimens without damaging them severely, because the electron beam would heat up the wet sample upon collision and more or less burn it. Moreover, biological specimens always were "thick" samples, and the use of the EM was more or less out of the question. The EM was better suited to the material sciences for scanning surfaces and very thin samples. Another competitor for microscopy at a sub-micron level was the newly invented Scanning Tunnelling Microscope (STM), for which the developers

Gerd Binnig and Heinrich Rohrer received the Nobel Prize in 1986. The stunning images produced with these new instruments became known extremely fast. For instance, the image of the IBM logo formed by 35 xenon atoms on a nickel surface, recorded by a variant of the STM, the Atomic Force Microscope (AFM), became an international sensation and an iconic representation of the instrument.

The TXM never produced an image of this quality, which is not to say that the images were not very impressive. On the contrary, in the mid-nineties, a joint effort of physicists and biologists at the Advanced Light Source, a synchrotron storage facility on the campus of the University of California, Berkeley (USA), managed to record a series of images illustrating the deterioration process of red blood cells that were infected with malaria. The microscope used there, the XM-1, was a sister setup of the Göttingen TXM developed by Schmahl, and parts of it had been produced by Schmahl's group. The lead physicist, Werner Meyer-Ilse, was an alumnus of the University of Göttingen and trained in Schmahl's group. The image series showed how the heavier elements (in the case of red blood cells' iron) seemed to form a trail towards the centre of the cell's body and finally form a clot of iron in the cell, which by then was no longer a viable entity. Although the series showed the effects that this particular disease had on red blood cells, the new insight was still not enough to convince the lead biologist in the experiment. The collaboration of biologists and physicists came to an abrupt end shortly after that. Biologist Cathy Magowan left science and joined the University of California administration services; physicist John Brown suffered from an infarction; and physicist Meyer-Ilse was killed in a car accident during a trip to his home country.[18]

The German branch of the TXM scientists, then already permanently stationed at BESSY and later on BESSY II – the Berliner Elektronenspeicherring-Gesellschaft für Synchrotronstrahlung (Berlin Electron Storage Ring Society for Synchrotron Radiation) – in Berlin, revived the idea of imaging biological samples with the TXM shortly after 2000, even though it was obvious that the TXM was also of good use in the material sciences.[19] Led by Günter Schneider, the BESSY II group imaged biological specimens and produced a three-dimensional animation of a plant cell (see *Fig. 7.6*). Unlike the image series from Berkeley (see *Fig. 7.5*) where the biology department was part of the research team, the image series made at BESSY II was rather driven by the

18 Cathleen Magowan, interview with Andreas Junk, Oakland, CA, 25 January 2005.

19 Peter Fischer *et al.*, "Imaging of Magnetic Domains with the X-Ray Microscope at BESSY Using X-Ray Magnetic Circular Dichroism", *Zeitschrift für Physik B* 101, no. 3 (1996), pp. 313–316.

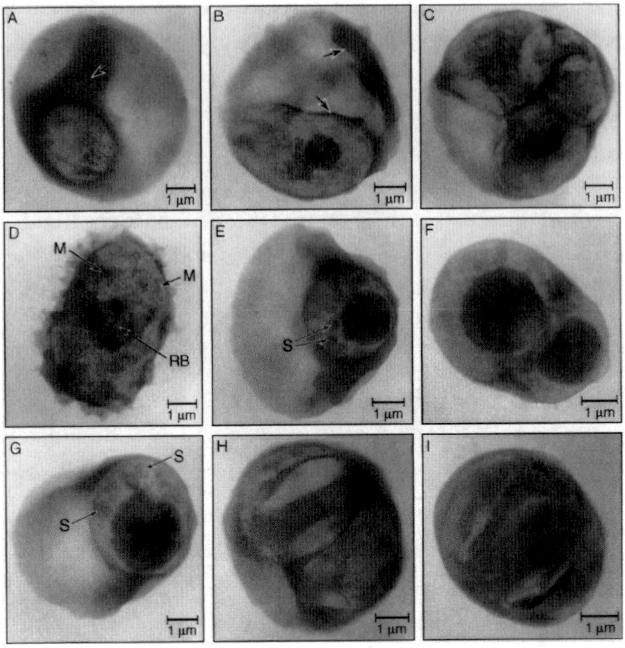

FIGURE 7.5 The series of deteriorating red cells shows how the superficial and intracellular structures are reorganised after a malaria infection. The increasingly darker zones indicate a major presence of iron, which absorbs the x-rays used most effectively. Reprinted from Cathleen Magowan *et al.*, "Intracellular Structures of Normal and Aberrant *Plasmodium Falciparum* Malaria Parasites Imaged by Soft X-Ray Microscopy," *Proceedings of the National Academy of Sciences* 94, no. 12 (June 1997), pp. 6222–6227, fig. 2
WITH PERMISSION FROM THE NATIONAL ACADEMY OF SCIENCES

resident physicists. The work was intended not only to show contrast between the absorbing specimen and the transparent background, which a TXM could provide even without a contrast agent. The images were also taken from different angles and computed into a tomographic image (see *Fig. 7.6*, right), so that a high-resolution, three-dimensional model of a plant cell was obtained.

The success in tomographically imaging the cell was somewhat guaranteed by enhanced preparation techniques and the necessary computing power. The images of biological specimens, of course, did not show living cells or still viable tissue. The preparation involved the shock freezing of the sample with liquid nitrogen in order to fix the specimen and increase the signal-to-noise ratio in the recording. When in a frozen state, there is a lot less scattering of radiation,

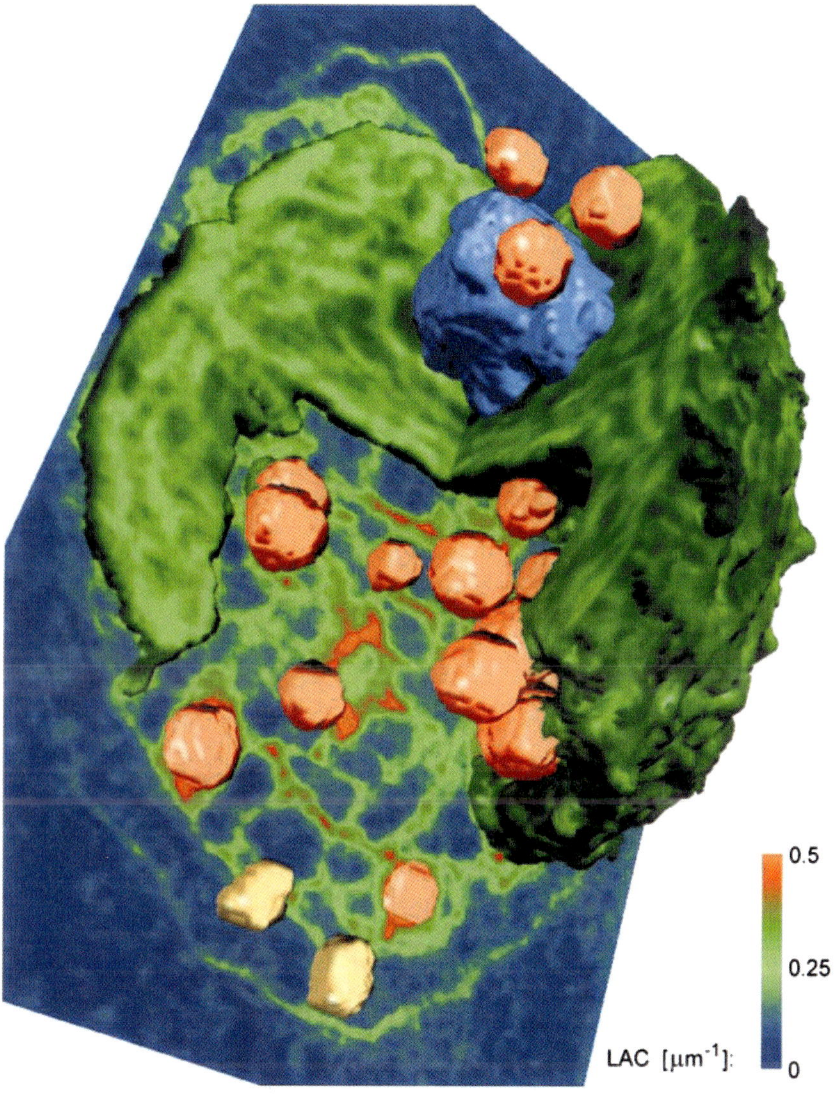

FIGURE 7.6 Tomographic "slice" of a plant cell (*left*), which together with other slices was computed into the cell's tomogram (*right*). Reprinted from D. Weiß *et al.*, "Computed Tomography of Cryogenic Biological Specimens Based on X-Ray Microscopic Images", *Ultramicroscopy* 84, nos. 3–4 (2000), pp. 185–197

and therefore less opportunity for the sample to be heated and destroyed. The output of TXM experiments showed the high grade of control the physicists had over their samples, but the results were seemingly not noticed outside the community of x-ray microscopists.

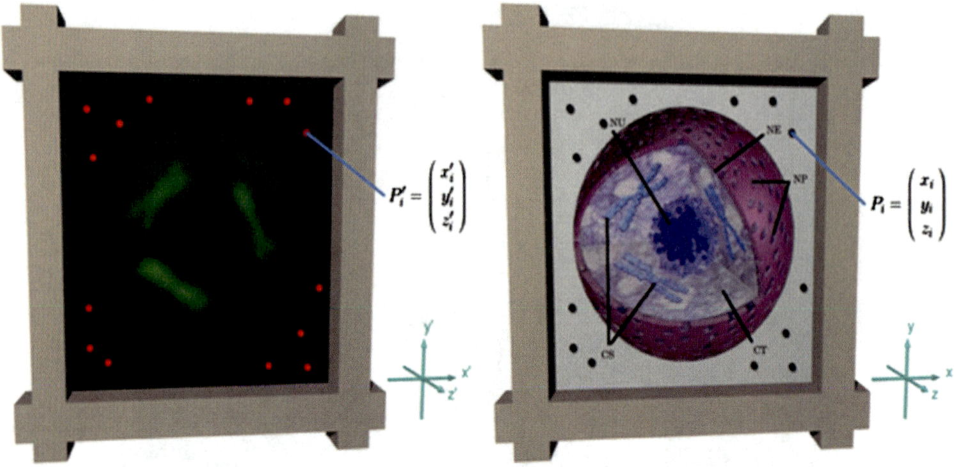

FIGURE 7.7 An idea for a TXM icon: the alignment of fluorescence microscopy (*left*) with the red markers for the orientation and the TXM image (*right*) with the detailed information about the cells content and arrangement. Reprinted from Gerd Schneider *et al.*, "Cryo X-Ray Microscope with Flat Sample Geometry for Correlative Fluorescence and Nanoscale Tomographic Imaging", *Journal of Structural Biology* 177, no. 2 (2012), pp. 212–223
WITH PERMISSION FROM ELSEVIER

The BESSY group then planned to demonstrate results with an iconic image whose characteristics would not only represent the actual scientific result but also that this result was obtained with a TXM. The idea of what this iconic image would have looked like is shown in *Figure 7.7*, which compares an image (on the right) made by TXM microscopy with an image (on the left) by fluorescence microscopy, a technique widely used in the life sciences. (*Fig. 7.7*) Showing the same specimen, the images should have provided very specific, singular, and also complementing information, indicated by the arrows in both. Side by side, the paired images showed viewers the information that could be gained with a TXM. Although the promotion of TXM microscopy in this way might have been successful, no publications presenting results in this style are known to me.

5 Conclusions

My evaluation of the development of TXM shows that the TXM was a failure, even if it was a working instrument with many advantages. From my interviews, I understood that the strategic goal was to build a super microscope for biology. Although Schmahl managed to establish a new idea to make the optics necessary for this purpose and found ways to test the optics, build the instrument, and even advance his career, the troubles to stabilise the experimental setup took so long that instruments like the STM or AFM could take the stage before the TXM did. The formation of a new community of x-ray microscopists was a success as well, but this group did not include many biologists; the new community became just another subset of physicists, who usually stayed among themselves because of their high degree of specialisation. The TXM did not become known to the life sciences as an instrument *needed* to produce a particular kind of high-resolution image for the work of biologists. In addition, the TXM could not be brought into university laboratories in large numbers, and the number of working groups at universities in this field remained very small. The lack of a sufficiently small and compact laboratory source of x-rays meant that researchers had to go to storage-ring facilities and compete for experimental time slots there. The promotion of the instrument was difficult as well because competing microscopes like the STM had produced spectacular images illustrating scientific results and the instruments themselves, whereas the TXM did not.

CHAPTER 8

From Complex to Simple: the Example of *Qibla*-Indicators

Razieh S. Mousavi

In the early modern period, the Safavids (r. 907–1148 AH/1501–1736 CE) founded an empire mainly in an area that is now Iran. The Safavids, who came from a religious order with military and political aspirations, established Twelver Shiʿi Islam as the state religion and consequently an identity that enabled them to unify their realm and counter neighbouring Sunni powers. To promote religious faith, the Safavids welcomed the migration of scholars from outside Iran, especially Lebanon. These legal specialists helped the new dynasty bolster its political legitimacy and propagate Shiʿi beliefs and rituals.[1]

Before the Safavid sovereignty, mathematicians were traditionally responsible for the scientific needs of religious rituals, including the determination of the times for praying and the observation of the crescent Moon for the beginning of Ramadan. Moreover, according to the Qur'an, Muslims must face the Sacred Mosque (*al-masjid al-ḥarām*) in Mecca when they do their daily obligatory prayers. Finding the praying direction to Mecca, namely *qibla*, had also been supervised by mathematicians. However, the Safavids' religious policy and the incorporation of Shiʿi jurists into the state legal system reduced the former authority of mathematicians in these affairs. Regarding the *qibla* determination, legal scholars supported and promulgated folk astronomical practices that were simpler and more available to ordinary people than complicated mathematical methods.

ʿAlī b. al-Ḥusayn al-Karakī (d. 940 AH/1533 CE) was a leading Shiʿi jurist from the region Jabal ʿĀmil – now in Lebanon – who migrated to Iran following the invitation of the Safavid monarch. Al-Karakī achieved considerable power at the court of Shāh Ṭahmāsb (r. 930–984 AH/1524–1576 CE) who appointed him as "the Seal of the Legal Scholars" (*Khātam al-mujtahidīn*). Another major

1 Notable among studies on the power division in Safavid Iran and clerical association with the court are Devin J. Stewart, "Notes on the Migration of ʿĀmilī Scholars to Safavid Iran", *Journal of Near Eastern Studies* 55, no. 2 (1996), pp. 81–103; Rula Jurdi Abisaab, "Delivering Justice: The Monarch's ʿUrfi Courts and the Shariʿa in Safavid Iran", in Anver M. Emon, Rumee Ahmed (eds), *The Oxford Handbook of Islamic Law*, Oxford University Press, Oxford, 2018, pp. 511–537: esp. 517–520.

 | DOI:10.1163/9789004689107_010

position at the court at that time was *Ṣadr* or religious chancellor who had responsibility for coordinating royal orders and organising administrative and bureaucratic processes to expand and promote religious affairs.[2] Because of the exclusively high power of al-Karakī, there were many disagreements between him and those who were appointed as *Ṣadr*. The friction, however, reached its peak when Ghiyāth al-Dīn Manṣūr Dashtakī (d. 948 AH/1541 CE), the prominent Iranian mathematician and philosopher of the time, got the position.[3] Dashtakī was against some of al-Karakī's clerical opinions, but the most intense dispute between the two scholars stemmed from their differing approaches in determining *qibla*.

The change in the direction of praying was not as easy as the other reforms, since it entailed realigning mosques' niches, whose constructions had mainly been guided by mathematicians up to that time. Al-Karakī's instruction was not tolerated by Dashtakī, who openly argued against it. The intensity of the disagreement escalated to the extent that they finally debated on the floor of the king's Chamber. Not surprisingly, the king sided with al-Karakī, resulting in Dashtakī's expulsion from the court. The historical sources provide little detail of the clash,[4] but we are fortunate enough to have the writings of both on their favoured methods for finding *qibla*, which allows us to learn more about the dispute.

The disagreement between these two significant figures of the early Safavid time has already been studied from the socio-political and religious points of view.[5] But no extensive attempt was made to analyse the issue in relation to the

2 About the powers and duties of those who were appointed as *Ṣadr* in the administration of charitable endowments as a significant practice within Islamic society, see Nuzhat Ahmadi, "*Vaqf* in the Safavid Period", in Rudi Matthee (ed.), *The Safavid World*, Routledge, London and New York, 2022, pp. 374–386: 375–376.

3 A summary of the disagreements between al-Karakī and Dashtakī appears in Rula Jurdi Abisaab, "Karaki" in *Encyclopædia Iranica*, Trustees of Columbia University in the City of New York, New York, 2020. http://dx.doi.org/10.1163/2330-4804_EIRO_COM_10749 (accessed 3 August 2023).

4 For informative reviews (though brief) of historical accounts on the conflict see, among others, Reza Pourjavady, *Philosophy in Early Safavid Iran: Najm al-Dīn Maḥmūd al-Nayrīzī and His Writings*, Brill, Leiden, 2011, p. 28; Ahab Bdaiwi, "The Youth Who Defeated Aristotle: The Life and Thought of Dashtakī (d. 948/1541)", *Global Intellectual History*, DOI: 10.1080/23801883.2022.2163915, published online: 2 Feb 2023, pp. 11–12. A more detailed discussion is in my Master thesis (in Persian): Razieh S. Mousavi, "Risāla-yi Qibla-yi Ghiyāth al-Dīn Manṣūr Dashtakī", Master thesis, University of Tehran, 2015, pp. 62–67, 87–95.

5 See for instance, Andrew J. Newman, "The Myth of the Clerical Migration to Safawid Iran: Arab Shiite Opposition to ʿAlī al-Karakī and Safawid Shiism", *Die Welt des Islams* 33 (1993), pp. 66–112: 99–103.

history of science as of now.[6] Even though the battle was more of a monopolising competition between two courtiers under certain political circumstances, the scientific discussion in the foreground could have significantly influenced the priorities of the intellectual community.

In the present study, we will analyse how mathematicians struggled to restore their previous credit and responded to the criticism over the complexity of their methods to find *qibla*. We will see how they worked on the design and construction of new instruments, to make the determination of *qibla* accessible to a wider group of worshippers with different levels of education. The advantage of such instruments relied on the fact that the sophistication and abstraction of mathematical operations could be well masked behind the functionality and mechanism of a tangible object. Generations of mathematicians thus devised and refined diverse *qibla*-indicators, whose stories of failure and success are the focus of this research. We will follow the thread of reforming *qibla*-indicators along the history of their production and see how instrument makers analysed the market demands and upgraded their products by offering attractive instruments in sizes, shapes, simplified functions, and multi-purpose applications. This also opens a window to examine how the complex mathematical instruments for *qibla* determination, like astrolabes and quadrants, shifted progressively towards user-friendly tools, which worked literally as input-output calculators.

1 Rise against Complexity

The methods for determining *qibla* in al-Karakī's juridical text are primarily based on natural references, like specific stars, winds, and the annual position of the Sun.[7] These methods were not invented by al-Karakī himself, but he inherited them from a long tradition of folk knowledge, on which legal specialists mainly established religious rituals.[8] In parallel, mathematicians and

6 For a concise analysis by Sonja Brentjes, in which she maintains that the disagreement between the two scholars was due to their support of calculative and geometrical approaches to the *qibla* issue, see: Sonja Brentjes, "The Mathematical Sciences and Medicine in Safavid Iran", in Rudi Matthee (ed.), *The Safavid World*, Routledge, London and New York, 2022, pp. 428–446: 434.

7 ʿAlī b. al-Ḥusayn al-Karakī (al-Muḥaqqiq al-Thānī), *Jāmiʿ al-maqāṣid fī sharḥ al-qawāʿid*, Muʾassasat Āl al-Bayt, Qum (Iran), 1408–1415 AH/1988–1994 CE, 14 vols.: v. 2, pp. 54–59.

8 For a comprehensive account on simple techniques for finding *qibla* in the so-called folk astronomy and sacred geography, see David A. King, *World-Maps for Finding the Direction and Distance to Mecca: Innovation and Tradition in Islamic Science*, Brill, Leiden, 1999, pp. 48–55. For a general overview on Islamic folk astronomy, see Petra G. Schmidl,

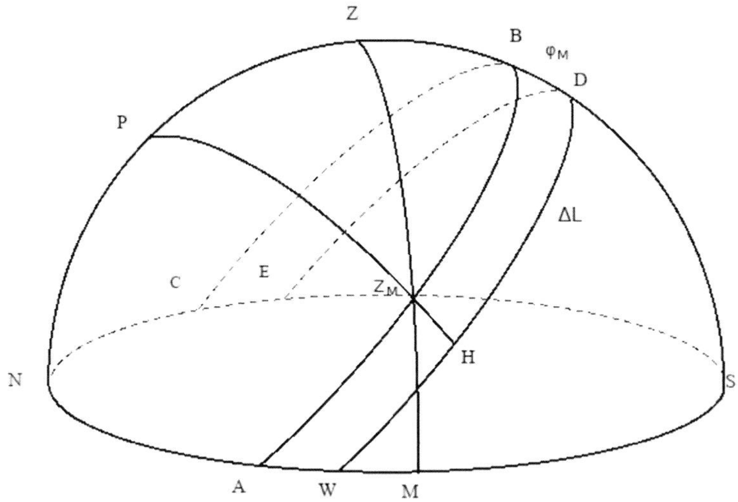

FIGURE 8.1 Finding *qibla* as a problem of spherical astronomy. In this diagram, *SEN* is the horizon circle of the city whose direction to Mecca is desired, *Z* is the zenith, *WDE* is the Equator, and *ABC* represents the parallel of Mecca whose zenith is shown as Z_M (in this regard, φ_M will be the latitude of Mecca and ΔL, the difference in longitude between Mecca and the longitude concerned). Accordingly, *qibla* can be determined as an angular magnitude by specifying the azimuth of Z_M on the horizon circle, namely the arc *SM*

astronomers had followed their own way of solving this question by using scientific definitions and techniques. They turned the question of *qibla* into a trigonometric problem, in which the celestial coordinates of the examined city and Mecca delineate *qibla* as an angle built upon latitudinal and longitudinal circles. The problem of finding *qibla* was thus adjusted into the framework of spherical astronomy, so that mathematicians could benefit from various tools to address the issue (*Fig. 8.1*).

This erudite tradition led to the creation of numerous texts replete with complex methods. Due to the difficulty in finding the solutions and the need for prerequisite knowledge, an exclusive space was set up for mathematicians to serve Islamic society and make them acquire more legitimacy in relation

"Islamic Folk Astronomy", in Clive L. N. Ruggles (ed.), *Handbook of Archaeoastronomy and Ethnoastronomy*, Springer, New York, 2014, pp. 1927–1934. https://link.springer.com/referenceworkentry/10.1007%2F978-1-4614-6141-8_199 (accessed 3 August 2023).

to their scientific practices.[9] Their books and instruments, however, were designed to appeal only to professionals and high-level literates and were totally inaccessible to laymen.[10]

Neglecting the public use of these methods provided Safavid legal scholars with a valid excuse to divert worshippers from referring to mathematicians on the *qibla* issue. It should be noted that the confrontation between legal and scientific approaches to this question had already started before this period, but it never went beyond scholarly discussions.[11] The religious rule of the Safavids offered a suitable platform for this long-term discrepancy to be employed as a political tool to magnify the authority of religious jurists over other elites.

Following the quarrel with al-Karakī, Dashtakī wrote a treatise on *qibla* in which he collected numerous mathematical methods introduced by his predecessors.[12] Although his text does not contain much new material from

9 According to D. A. King, from the ninth century onwards, Muslim scholars had approximate and accurate (simple vs. sophisticated) mathematical methods for finding *qibla* at their disposal, but they were not always consulted. He also argues that relying on mathematicians for religious timekeeping and *qibla* determination varied remarkably in different socio-cultural contexts in the medieval period which could have led to different stories of disagreements between scientific and legal methods. David A. King, *In Synchrony with the Heavens: Studies in Astronomical Timekeeping and Instrumentation in Medieval Islamic Civilization*, Brill, Leiden, 2004–2005, 3 vols.: v. 1, pp. 669–672. For an example of a local challenge in determining *qibla*, see Mònica Rius-Piniés, "Qibla in the Mediterranean", in Ruggles, *op. cit.* (n. 8), pp. 1687–1694. https://link.springer.com/referenceworkentry/10.1007/978-1-4614-6141-8_171 (accessed 3 August 2023).

10 A good example is the book on mathematical geography by the prominent Muslim scholar, al-Bīrūnī (362–*c.*440 AH/973–*c.*1048 CE), entitled *Taḥdīd nihāyāt al-amākin* ("The Determination of the Locations of Cities"), in which he sets the determination of *qibla* for a certain city as his ultimate goal. Complaining about folk methods in his introduction, al-Bīrūnī concludes that, given the significance of the direction of praying in the life of every Muslim, only accurate mathematical methods must be used to delineate it. Abū Rayḥān Muḥammad al-Bīrūnī and Muḥammad al-Ṭanjī (ed.), *Taḥdīd nihāyāt al-amākin li-taṣḥīḥ masāfāt al-masākin*, Doğuş Ltd. Şirketi Matbaası, Ankara, 1962, pp. 14–16.

11 For a thirteenth-century debate between Naṣīr al-Dīn Ṭūsī (d. 672 AH/1274 CE) and al-Muḥaqqiq al-Ḥillī (d. 676 AH/1277 CE), see Michael Cook, "Why Incline to the Left in Prayer? Sectarianism, Dialectic, and Archaeology in Imāmī Shīʿism", in Michael Cook, Najam Haider, Intisar Rabb, and Asma Sayeed (eds), *Law and Tradition in Classical Islamic Thought, Studies in Honor of Professor Hossein Modarressi*, Palgrave Macmillan, New York, 2013, pp. 99–124: 104–105.

12 For a detailed study of the mathematical methods in Dashtakī's book, see Mousavi, "Risāla", *op. cit.* (n. 4), pp. 185–249. D. A. King has conducted exhaustive studies on numerous scientific methods for finding *qibla* in Islamic societies. For instance, see King, *World-Maps, op. cit.* (n. 8), pp. 56–89. Moreover, on maps and instruments for finding *qibla*, see David A. King and Richard P. Lorch, "*Qibla* Charts, *Qibla* Maps, and Related Instruments", in J. B. Harley and David Woodward (eds.), *History of Cartography*, vol. 2,

the technical point of view, it is innovative in its structure and in the way it describes the existing methods. Dashtakī recognised the inapproachability of mathematicians' contributions as the core of the criticism, which he tried to dispel by retelling and reformulating already developed knowledge. Two main features make his work distinct from previous ones:

1) He classifies the existing mathematical methods into different categories according to the needed prerequisite knowledge, so that the reader can choose the more suitable method for him. This style is in remarkable contrast concerning most of the previous methods that were often part of a large complex work on mathematics or astronomy.
2) Unlike prior works, the methods are described in a simple, step-by-step way, without providing any proof or theoretical explanation.

The above points underline Dashtakī's attempt to present in a simpler and more accessible way the mathematical solutions proposed to find *qibla*. Among the various collected methods, which are primarily computational, geometrical, and observational (with different levels of approximation), only two instruments for finding *qibla* are mentioned: the astrolabe and the magnetic compass.

This helps us better to highlight the changes in practices that were to be introduced after Dashtakī, when diverse instruments for finding *qibla* were invented, produced, and largely used. We will see examples of such instruments later in the paper. Dashtakī's treatise also suggests that there was no instrument called *qibla*-indicator before this time. The astrolabe was actually a pre-Islamic astronomical instrument, on which graphical quadrants for finding *qibla* were occasionally inserted by Muslim designers. *Figure 8.2* shows an astrolabe of this kind held in the History of Science Museum in Oxford (inv. no. 45747). A family of curves are engraved on the back of astrolabe, representing the altitude of the Sun at the azimuth of the *qibla* for certain cities.[13] Moreover, by adjusting the position of the Sun on the astrolabe to the zenith of Mecca with respect to the horizon of a given city, the *qibla* can be found by determining the azimuth of the Sun in that setting.

As for the magnetic compass, however, Dashtakī introduces it as a "new" method for finding *qibla*. This does not mean that the compass was not known

---

book 1: *Cartography in the Traditional Islamic and South Asian Societies*, University of Chicago Press, Chicago and London, 1992, pp. 189–205.

13 Although the method was theoretically known in the ninth century, the earliest surviving example of its use is a Safavid astrolabe, and it remained popular from then onwards. Despite its visualised method for finding *qibla*, it was not as simple as expected. For more on these quadrants and their practical issues, see King, *World-Maps*, *op. cit.* (n. 8), pp. 186–193.

FIGURE 8.2 The upper-right quarter of the back of this astrolabe is a quadrant for finding the direction to Mecca from some desired cities. The horizontal axis is scaled, based on the months of a year. The curves indicate the yearly position of the Sun in specific cities whose names are marked on the circumference of the quadrant. The *qibla* argument of a city at a specific time of the year can be obtained through drawing a line between the centre of the circle to the intersection of the related curves

to Muslims before his time, as witnesses verify its use by navigators and in rare cases also for *qibla* determination.[14] Dashtakī's note rather implies that finding the direction to Mecca by means of a compass was not a common practice among mathematicians at his time. This may be due to their recognition of the difference between the geographical (celestial) or true North and the magnetic North.

At any rate, a growing number of instruments for finding *qibla* had compasses inserted after Dashtakī's treatise, which could offer ease of use as a significant quality expected from *qibla*-indicators. This change in scientific approach indicates that, after the court dispute, mathematicians paid more attention to the potential of user-friendly instruments in order to facilitate individual *qibla*-finding and attract more users.

Inspecting the devices invented by the following generations of Persian scholars reveals a line of endeavour to conceal unappealing complex methods behind handy elegant instruments. While the astrolabe was mainly used by versed scholars, *qibla*-indicators were rather designed to be used by ordinary people. In fact, enthusiasts for mathematical methods adopted the strategy of popularising their ideas through instrumentation, in order to continue their social services in the problem of *qibla* determination. From the sixteenth to the nineteenth centuries, Islamic society in Iran witnessed an explosion of *qibla*-indicators aimed at being used with less prior understanding of mathematics. We will follow the line of evolution of some of the devised instruments in a rough chronological arrangement in the next section. At the outset, it should be noted that rarely do we have surviving instructions or manuals for *qibla*-indicators. Therefore, the present study has been based mainly on the analysis of extant instruments, as well as on the underlying mathematics and their functions.

2 *Qibla*-Indicators in Transition

Much has already been published on the mathematical and technical aspects of Safavid *qibla*-indicators, but the exploration of the thread of continuity between the surviving instruments, as this present paper aims to accomplish, remains infrequent in literature. The demand for simpler methods provided

---

14 Information about the use of compasses for finding *qibla* in pre-Safavid time is sparse. For two of the earliest known pieces of evidence, see Petra G. Schmidl, "Two Early Arabic Sources on the Magnetic Compass", *Journal of Arabic and Islamic Studies* 1 (1997–1998), pp. 81–132: 86–88.

the ground to draw scholarly attention from complex, accurate methods to approximate, algorithmic procedures. In other words, approximation, hitherto unfavoured by mathematicians, progressively turned into a normative way of determining *qibla*.

For centuries, multiple lists of *qibla* values were prepared by Muslim scholars for hundreds of localities through diverse computational methods. By means of these values, commonly in tabular formats, it was indeed very easy to determine *qibla* in various locations. To use a table, one only needed to find the proper cell in a two-columned table of place names and *qibla* values, as well as to know the north-south direction of the examined town in advance. Although all the requirements were available to devise simple *qibla*-indicators by combining the *qibla* values with magnetic compasses, this action was delayed until the time of the Safavids, when the virtue of simple performance replaced the concern of mathematicians for high precision. Thereby, the compass, regardless of its deviation from true geographical directions, became increasingly popular in creating new *qibla* instruments.

In late Safavid times and afterwards, *qibla*-indicators comprising a compass and a map or gazetteer of *qibla* values for major cities were in widespread use. While earlier models were designed to "find" *qibla* values, map-based models were produced to "display" already computed values. We distinguish these two types of *qibla*-indicators by referring to them as "*qibla*-finders" and "*qibla*-pointers", respectively. A major goal of pre-Safavid mathematicians had focused on the improvement of technical methods to find *qibla*, whereas Dashtakī and his descendants paid more attention to presentation techniques and algorithmic functions. In this context, computational methods lost favour over time and visual/geometrical tools gained more popularity.

In parallel, artistic works received special attention under the reign of the Safavids. Politically, it was due to the Safavids' strong demand for a centralised state supported by a skilled class of craftsmen. The development of art (including metalwork) reached new heights of perfection in this period, which influenced the design and quality of scientific instruments as well.[15] This fertile ground might have facilitated the cooperation between mathematicians and instrument makers to design and fabricate new *qibla*-indicators. However,

---

15 For an informative summary on Safavid art and its correspondence with political factors, see A. Welch, "Art in Iran, ix. Safavid to Qajar Periods" in *Encyclopædia Iranica*, v. 2, pp. 620–627. https://iranicaonline.org/articles/art-in-iran-ix-safavid (accessed on 4 August 2023).

some of the projects did not go beyond written works, whereas some others were produced in large numbers.[16]

A remarkable innovation was made by a famous Safavid mathematician, Muḥammad Bāqir Yazdī (d. 1056 AH/1647 CE), who described it in a treatise in Persian, entitled *Maṭlaʿ al-anwār*, which was the name of his newly designed instrument. From the surviving text, we learn that Yazdī's new astronomical compendium was fabricated at least once, to be dedicated to a Safavid king. *Figure 8.3* provides a schematic picture of this device, which is composed of two perpendicular plates.[17] In his introduction, Yazdī declares that the instrument, despite its numerous applications as an astrolabe, was invented in an effort to determine the *qibla* direction. This assertion documents once again the centrality and significance of this question at Yazdī's time, in contrast to a few centuries earlier, when this was only a by-product of mathematical and astronomical instruments, including astrolabes. Yazdī suggests three techniques for finding *qibla* by utilising the quadrant scale of the horizontal plate and its rotating ruler. None of the methods is technically new;[18] the asset of the instrument is, however, its capacity of turning the computational and geometrical procedures of earlier methods into instrumental functions, to exempt the user from dealing with advanced mathematics. In fact, the methods are broken down into a set of physical actions via moving the graduated ruler to a certain degree in a certain direction. We recognise a tendency towards "mechanising"

16 In his well-known Persian treatise, Raḍiyy al-Dīn Qazwīnī (d. *c.*1095 AH/1684 CE) explains how he recalculated *qibla* values through devising a new instrument, similar in construction to a geographical globe. At the end of the treatise, the attained *qibla* arguments for different localities are recorded as a list. Although dozens of copies of this treatise are extant, there is no surviving instrument of this type, which implies that Qazwīnī's instrument remained as a one-off model. This example signals that there was a reluctance to use "*qibla*-finders", while computed values were more desirable. For a mathematical analysis and a visual reconstruction of the instrument, see Majid Heydari Delgarm, "A Previously-Unknown Iranian Treatise on a Terrestrial Globe", *Studies in History and Philosophy of Science* 95 (2022), pp. 204–214.

17 I gave a talk on this instrument in September 2016 at the 35th Scientific Instrument Symposium in Istanbul, under the title "An Iranian Quadrant for Finding *Qibla*", in which I explained Yazdī's three mathematical methods to find *qibla* using the instrument. I was recently informed that the whole treatise of Yazdī has been studied as a project for a Master's degree at the University of Tehran.

18 The first method is the so-called "method of *zīj*es" which is basically computational (for a brief explanation, see the caption of *Fig. 8.3*), while the other two are geometrical solutions. For an informative overview on the technical aspects of these methods in Islamic scientific literature, see King, *World-Maps*, *op. cit.* (n. 8), pp. 61–64.

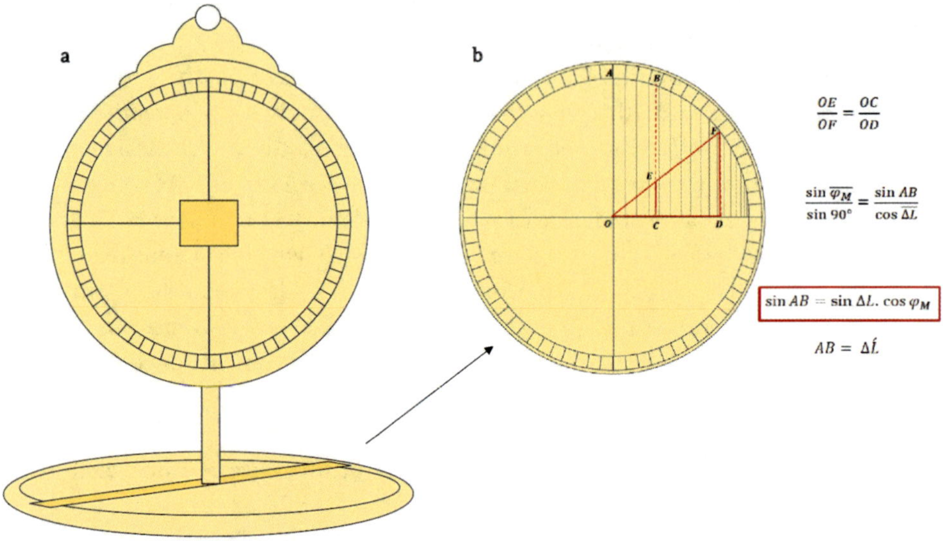

FIGURE 8.3 A rough schema of *Maṭlaʿ al-anwār* devised by Muḥammad Bāqir Yazdī (a). Alongside numerous applications, the horizontal plate (b) offers three different methods for finding *qibla* of which the "method of *zījes*" is illustrated here. The trigonometric relations show the computational steps attained through moving the scaled ruler attached at the centre of the plate in two steps. Note that the mathematical notations follow conventions in *Fig. 8.1*.

multi-step mathematical operations to facilitate *qibla* delineation.[19] The text seems to be aimed primarily at non-professional practitioners since it does not mention even the name of the underlying mathematical methods to his audience, let alone the proofs.

Notwithstanding its simple practice, the structure of Yazdī's *qibla*-finder was not suitable for mass production, as one always needed the instruction manual to follow the rules of operation. Therefore, Yazdī's *qibla*-indicator was not a good option for users unfamiliar with technical instruments. As we will see below, these aspects were improved in the design of later *qibla*-indicators. Producers actually became more and more drawn to appearance, affordability, efficiency, and simplicity. Professional applications such as astrolabic functions

19 In using the term "mechanisation of mathematics", I am inspired by Jane Wess's paper presented at the British Society for the History of Mathematics conference at Rewley House, Oxford, in 2008. Analysing the structure and function of several instruments in the Science Museum in London, she has shown how their inventors were aimed at avoiding mental computations. Jane Wess, "Avoiding Arithmetic, or the Material Culture of Not Learning Mathematics", BSHM *Bulletin: Journal of the British Society for the History of Mathematics* 27, no. 2 (1 July 2012), pp. 82–106.

were no longer a matter of concern, and a sundial (including a folding gnomon) was the only option that accompanied *qibla*-indicators to offer an attractive combination for worshippers who also needed to determine prayer times.

From this point on, compasses started playing a pivotal role in *qibla*-indicators. As we will see, most of the extant devices bear a compass which facilitates finding the (approximate) north-south direction of the desired locality without the need for any observation. *Figure 8.4* shows a rectangular instrument for finding *qibla* based on the functionality of a compass hinged to its top, missing today. This *qibla*-indicator, preserved at the History of Science Museum in Oxford (inv. no. 48472), is unfortunately unsigned and undated but is estimated to have been made in the eighteenth century. A gazetteer displaying the deviation angle (*inḥirāf al-qibla*) of fifty-nine cities is engraved on its back and the graphical direction to *qibla* for Isfahan (in central Iran) is on its front; these suggest that the instrument was meant for citizens of Isfahan, who might travel to the places listed in the gazetteer. Two other instruments on the front side are a horizontal pin-gnomon dial engraved with hour-lines and the directions to some Shiʿi holy cities in graphical form in the lower half.[20] A number of extant examples of rectangular *qibla*-indicators, in brass or wood, most of which are accompanied by sundials and gazetteers for multiple localities exist today in various collections.[21]

Concerns about universality are noticeable in several *qibla*-indicators. A developed model, better known as the "Safavid world-map", in which Mecca is located at the centre of the instrument's plate, has already been the subject of a number of extensive studies by historians of science.[22] A rotating ruler

20 Referring to a similar extant instrument, D. A. King has suggested a well-known Safavid instrument maker, ʿAbd al-Aʾimma, as the producer of this *qibla*-indicator. King, *World-Maps, op. cit.* (n. 8), p. 118.

21 For another interesting *qibla*-indicator in a rectangular form, see King, *World-Maps, op. cit.* (n. 8), pp. 95–98. It is signed in 1246 AH/1830 CE, and its accompanying inscription bears the names of both a designer and a copyist, as well as a specific name assigned to the design, namely *Jām-i gītī-numā* ("Glass of the Universe"). A treatise explaining the invention story of the instrument has survived, and is investigated in detail in a paper of mine: Razieh S. Mousavi, "Jām-e Gītī-namā: A Late Iranian Qibla-Indicator", *Tarikh-e Elm (Iranian Journal for the History of Science)* 16 (2016), pp. 17–40.

22 For a series of discussions on the origin of this instrument, see respectively: King, *World-Maps, op. cit.* (n. 8), esp. pp. 365–371; Elly Dekker, "Cartographic Grids from Iran: An Early Version of the Retro-Azimuthal Orthographic Projection?", *The Cartographical Journal* 37, no. 2 (2000), pp. 109–116; Jan P. Hogendijk, "Het mysterie van de Mekkawijzers van Isfahan", *Nieuwe Wiskrant* 22, no. 2 (2002), pp. 4–11; and King, *In Synchrony, op. cit.* (n. 9), pp. 825–846.

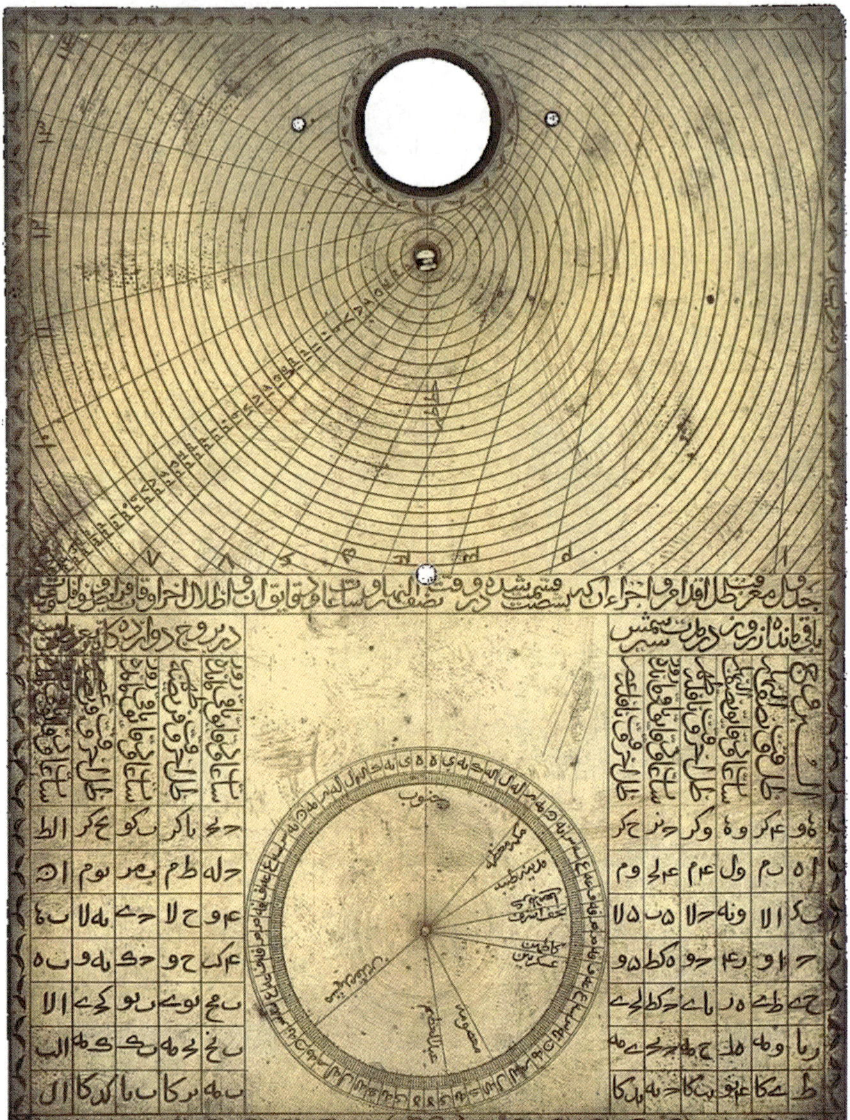

FIGURE 8.4 A Persian, rectangular *qibla*-indicator, now with empty housings for a hinged compass (*on top*) and a gnomon (*at the centre*). Unsigned and undated, possibly eighteenth century

attached at the position of Mecca points to the *qibla* when it traverses over the name of a desired city, thereby making the finding of *qibla* easy. The simple performance and visual impression of the instrument do not reveal the complexity of the mathematical method behind its design. In this instrument, no city is preferred over another, which makes it equally attractive to residents of various cities. Moreover, a brief Persian instruction on how to use the qibla-indicator and its accompanying sundial is engraved on the world map in a three-versed poem, making the operation easy to memorise. As a result, the user no longer needs to carry or refer to a manual. Perhaps because all the necessary information has been disguised by the physicality of the instrument, which could easily be copied by other instrument makers, no text containing its underlying mathematical methods has come down to us and its origin remains obscure.

A simplified version of the above instrument is also kept at the History of Science Museum in Oxford (inv. no. 43645), whose wooden material has made it lighter and cheaper, but less robust (*Fig. 8.5*). The small dimensions of the device would fit well into a pocket, while its small ruler (or index) is easy to dismantle and thus is prone to getting lost, as in the extant example. A detailed instruction is written on its back, and the same poem of the previous model appears on its rim, summarising the instruction. A specific name has been assigned to the device by the unknown maker, *Ṣafḥa-yi jahān-numā* ("Plate of the World") and, in prose, we read that since some cities did not fit on the front plate, they are listed in a circular gazetteer on the back. Accordingly, it seems that the maker felt obliged to include a certain number of cities on his instrument.[23]

The advantages of previous devices are gathered in a late *qibla*-indicator shown in *Fig. 8.6*, which is again housed in the History of Science Museum in Oxford (inv. no. 34566). It has the form of a magnetic compass fitted with a sundial. According to a short poem engraved on its outer rim, the device bears the specific name as *Āyina-yi gītī-numā* ("Mirror of the Universe"), which appears to function as a symbolic testament to the ownership of the underlying idea.[24] This small brass *qibla*-indicator, which is only a few steps

23 Further information on this grid-less Mecca-centred instrument can be found in King, *World-Maps*, *op. cit.* (n. 8), pp. 96–97, 100, 359.

24 It is worth noting that the names of the last three *qibla*-indicators cited here bear close resemblance (see footnote 21; caption of *Fig. 8.5*). They all are metaphorical of the projection of the world on a surface which indicates their visualisation value. Moreover, the instruction is engraved on the tool as a piece of poetry in all cases. These similarities along with step-by-step developments imply a tight competition in capturing the market and absorbing public interest. Also, the juxtaposition of craftsmanship and poetry emphasises the artistic and aesthetics values of *qibla*-indicators. I have a forthcoming paper in which I collect, analyse, and compare the Persian poems engraved on *qibla*-indicators

FIGURE 8.5 *Ṣafḥa-yi jahān-numā* ("Plate of the World"): a Mecca-centred *qibla*-indicator accompanying a horizontal pin-gnomon sundial. Persian lines of instruction and a gazetteer of towns appear on the back of the instrument

remote from modern versions, was widely copied in nineteenth-century Iran.[25] Its users did not need to have any clue of the mathematical relations

---

to highlight how understanding their literary aspects can be indicative of mathematical techniques.

25 The special poem engraved on the rim of these *qibla*-indicators has allowed me to recognise many tools of this type. For instance, some surviving examples at the History of Science Museum in Oxford are under the following inventory numbers: 34566 (*Fig. 8.6*), 50254, 50596, and 53791; their images are available online at the museum's

FIGURE 8.6 A late *qibla*-indicator fitted with a compass and a horizontal pin-gnomon sundial. A gazetteer of numerous *qibla* arguments is engraved on its back, and on the inner and outer sides of its lid
HISTORY OF SCIENCE MUSEUM, UNIVERSITY OF OXFORD, 34566

employed in the production, nor any visual understanding of *qibla* as a question in spherical astronomy. It only takes rule-following actions to perform the algorithmic steps that are presented in a poem. Thus, the user was only required to be able to line the device along the north-south direction, read the *qibla* angle from the gazetteer, and mark it on the graduated plate. Moreover, the instrument was smaller in size than previous models, which made it easier to carry in the pocket while travelling.

3 Concluding Remarks

The examples of *qibla*-indicators we just examined show how mathematicians and instrument makers shifted their attention to the advantage of the users' domain knowledge by focusing on practical aspects. After being criticised for being unwelcoming and inattentive to the general public and for neglecting the level of simplicity expected by legal scholars, instrument makers translated theoretical methods into graphical, physical, operational designs, in the way that mathematical and cartographical data can be represented in an easy-to-understand map to suit the needs of ordinary people. The transitional stage between Safavid *qibla*-finders and *qibla*-pointers discloses increasing familiarity of instrument designers with operational issues. They incorporated numerous techniques of simplification and popularisation, primarily directed at the consumer market, by considering user experiences and gathering feedback. To make applications easy and popular, mathematical methods in algorithmic language were conveyed through the friendliness and mnemonic features of poetry. Through this journey, underlying theories remained almost untouched from the point of view of precision and technical innovation,

website. Also, the registration number of another example at the British Museum is as follows: 1890,0315.4; image available online at https://www.britishmuseum.org/collection/object/W_1890-0315-4 (accessed 7 August 2023). Following my presentation of a draft of this paper at the 39th Scientific Instrument Symposium in London (virtual), Peggy Kidwell kindly informed me of another *qibla*-indicator of this type at the National Museum of American History with the inventory number MA.322416, which bears the same piece of poem; image available at https://www.si.edu/object/portable-horizontal-sundial-and-quibla-indicator:nmah_1215160 (accessed 7 August 2023). There seem to be many examples remaining to receive attention. Without referring to their identical poetry, D. A. King has introduced the technicality of this type of *qibla*-indicator as a popular example in pre-modern Iran (King, *World-Maps*, *op. cit.* (n. 8), pp. 121–123). It would be of great value to conduct a more detailed investigation in order to answer the question whether all of the extant *qibla*-indicators with these technical characteristics have the same artistic hallmarks, especially the piece of poetry.

while functional and presentational qualities underwent a huge development, aimed at replacing abstract procedures with visual interfaces.[26] As we saw, mathematicians subverted the political threat of their time by sacrificing part of the accuracy for the purpose of applicability. Indeed, the new value system to avoid failure resulted in an epistemic realignment under which certain performance expectations were seen as normative considerations. Although it was a conscious move in a new direction by those close to the debate time, subsequent generations of instrument makers seemed to have involuntarily contributed to the new standards.

The result of this study sheds light on what failure of an instrument might be. It highlights that a successful instrument does not always need the highest degrees of exactitude and sophistication. The ability to meet or not meet a target group's expectations might be regarded differently where and when the context has varied. Moreover, the expansion and diversity of users may change whether an instrument is assessed as a failure or success. Such considerations emphasise the importance of embedding studies of scientific instruments into their historical and cultural backgrounds, and also raise questions about how science may have developed and responded to the needs of society beyond and even opposite to its intrinsic values. The findings of this research also challenge previous studies that have tried to understand the widespread use and development of *qibla*-indicators in Safavid Iran by examining their similarities with European instruments, with little attention to cultural coordinates.[27] I would like to highlight that imported ideas can rarely bear fruit without fertile soil and proper conditions.

From this examination, it has been shown that an analysis of the forms of mathematical procedures incorporated into scientific instruments in order to serve the public can open new avenues up to further understanding of practical knowledge. Moreover, insights into the developmental stages of *qibla*-indicators towards easier function, and their integration with literary and aesthetic interests permit us to see a continual interchange between society and science. Accordingly, the commitment to improve accuracy in mathematical techniques was challenged, and more attention was directed to public demands.

---

26 The concern of balance and equalisation of values is broadly and inspiringly studied by Joel Kaye in *A History of Balance, 1250–1375: The Emergence of a New Model of Equilibrium and Its Impact on Thought*, Cambridge University Press, Cambridge, 2014.

27 For a recent study of this kind, see Anthony Turner, "A Mingling of Traditions, Aspects of Dialling in Islam" in Neil Brown, Silke Ackermann, and Feza Günergun (eds.), *Scientific Instruments between East and West*, Brill, Leiden and Boston, 2019, pp. 108–121: 117–121.

Another point that deserves scrutiny on a broader basis is the approach of legal scholars to the introduction of new and diverse *qibla*-indicators. Historical sources show that al-Karakī's rejection of mathematical methods in determining *qibla* had opponents also among the religious elites,[28] which probably served the rapid legitimacy of these tools in Islamic society. In passing, the present paper may also contribute to more exhaustive studies proving the groundlessness of the claim that the Safavid era was a period of scientific decline.[29] The dynamics underlined in this research suggest that the Safavid world and its aftermath witnessed a remarkable variety of creative minds.

Let us underline that this research does not claim to have investigated all types of surviving *qibla*-indicators; it is rather, a modest attempt to study and compare several examples from different sources, to provide a snapshot of how socio-political settings influenced ways of thinking and practice in modelling instruments for determining *qibla*. I hope this paper will be an incentive for researchers to study *qibla*-indicators in conjunction with various aspects of mathematical sciences within Islamic societies, contributing to a more comprehensive understanding of this rich tradition.

Acknowledgements

I am grateful to Stephen Johnson and Senthil Babu whose comments and suggestions were of great value to this paper.

28 One of the first and foremost objections was raised by Ḥusayn b. ʿAbd al-Ṣamad al-Ḥārithī (d. 984 AH/1576 CE), whose polymath son, better known as Shaykh Bahāʾī (d. 1031 AH/1622 CE), served the Safavids in the same position as al-Karakī. For a copy of al-Ḥārithī's treatise on refuting al-Karakī's opinions on determination of *qibla* in Iran and Iraq, see: Tehran, Majlis Library, MS 15616/2, fols. 63r–74r.

29 For a useful study that rebuts this misinterpretation by elucidating mathematical sciences under Safavids, see Sonja Brentjes, "The Mathematical Sciences in the Safavid Empire: Questions and Perspectives", in D. Hermann and F. Speziale (eds.), *Muslim Cultures in the Indo-Iranian World during the Early-Modern and Modern Periods*, Klaus Schwarz Verlag, Berlin, 2010, pp. 325–402. For a more recent and comprehensive study, see: Nahyan Fancy *et al.*, "Current Debates and Emerging Trends in the History of Science in Premodern Islamicate Societies" *History of Science* 61, no. 2 (2023), pp. 123–178.

CHAPTER 9

Pedro Nunes, Tycho Brahe, and Orazio Borgondio: Two Failed Schemes to Read Angular Scales?

Luís Tirapicos

This essay addresses two historical cases of apparent scientific instrument failure linked with the division and reading of angular scales. They are both, in some way, connected to the Portuguese milieu, a main focus of my historical research efforts, but clearly immersed in a wider European context.

The narrative starts with a discussion of the nonius scale, proposed by Pedro Nunes (1502–1578) in 1542. Although the nonius was never widely adopted, it had a long history spanning the seventeenth century and crossed the paths of Tycho Brahe (1546–1601), Christopher Clavius (1538–1612), and Andrés Garcia de Céspedes (*c.*1545–1611). By contrast to this well-known instance, my second story deals with a rather obscure device proposed by the chair of mathematics of the Roman College, Orazio Borgondio (1675–1741), in 1720, and used by the Jesuits in Lisbon in the latter half of the 1720s.

Pedro Nunes was a remarkable sixteenth-century Portuguese mathematician and cosmographer. Scholars have long debated his possible Jewish origins but the available evidence is, to say the least, contradictory and uncertain.[1] What is well documented about Pedro Nunes is that he was born in 1502, in Alcácer do Sal, an old village southeast of Lisbon. Nunes studied in the 1520s at the University of Salamanca, becoming a Bachelor in medicine around 1526. After his return to Portugal in 1529, he was named Royal Cosmographer by King João III (r. 1521–1557), inaugurating a lineage of Portuguese official cosmographers, and started a career at court that would last until his death – namely teaching mathematical and philosophical matters to the brothers of João III: Dom Luís (1506–1555), Dom Henrique (1512–1580), and Dom Duarte (1515–1540). In 1547, Nunes was designated the Royal Chief-Cosmographer, the first time this title was given in the Portuguese kingdom.

In the early 1530s, Pedro Nunes lectured on logic, metaphysics, and philosophy at the University of Lisbon, where he obtained a doctorate in medicine in March 1532. After the Portuguese university was moved to Coimbra in 1537,

1 The early biography of Pedro Nunes and his formative years as a mathematician are discussed in Henrique Leitão, "Para uma biografia de Pedro Nunes: O surgimento de um matemático, 1502–1542", *Cadernos de Estudos Sefarditas* 3 (2003), pp. 45–82.

 | DOI:10.1163/9789004689107_011

Pedro Nunes became a teacher there. He worked there between 1544 and 1562, with several interruptions due to his being called to Lisbon in his capacity as the royal cosmographer in order to solve nautical questions. Although there remains some evidence of Nunes's direct collaboration with navigators, cartographers, and pilots, his published works denote a primarily theoretical and mathematical approach to navigation, leading several historians to see him as the founder, or one of the founders, of navigational science.[2] Pedro Nunes died in Coimbra in 1578.

In 1542, in his book *De Crepusculis* (*Fig. 9.1*), Nunes discussed with great generality the mathematical problem of the duration of twilight during the year and in a given latitude, and proposed the introduction of additional divisions on an astrolabe to permit a more accurate reading of altitude angles. This device was later known as the nonius (the Latin form of his name, "Nunes").[3] According to him, there would be added inside the usual quadrant divided into 90 degrees, another arc dividing the quarter-circle into 89 equal parts. Inside this 89-part arc, another quarter-circle of 88 equal parts would be added, and so on to the innermost arc, which had 46 equal divisions. The observer had only to recognise to which of the marks the alidade was aligned. For an alignment with the ath mark of an arc divided into n parts, the resulting angle was obtained dividing a by n and multiplying by 90°.[4] The idea of the nonius did not appear by chance but, as stated by Nunes in his work *De arte atque ratione navigandi libri duo* (1573), was suggested to him by a passage in Ptolemy's *Almagest* where the Alexandrian refers to the arc of meridian between the

2 See Bruno J. de Almeida, "A Influência da obra de Pedro Nunes na náutica dos séculos XVI e XVII: um estudo de transmissão de conhecimento", doctoral dissertation in the History and Philosophy of Science, University of Lisbon, 2011; and Henrique Leitão, "Sobre a difusão europeia da obra de Pedro Nunes", *Oceanos* 49 (2002), pp. 110–128. This evaluation results from the fact that Nunes was the first, in two studies published in his *Tratato da Sphera* (1537), to apply mathematical techniques to nautical problems, but also from the impact and diffusion of Nunes's works in Europe, in particular among cartographers and the authors of nautical treatises.

3 For a detailed study of Nunes's *De Crepusculis*, see Eberhard Knobloch, "Nunes' *Book on Twilights*", in L. T. de Campos, H. Leitão, and J. F. Queiró (eds.), *International Conference Petri Nonii Salaciensis Opera. Proceedings*, Faculdade de Ciências da Universidade de Lisboa, Lisbon, 2003, pp. 113–140. On the historical dimension of the nonius scale see António E. dos Reis, "O Nónio de Pedro Nunes", *Gazeta de Matémática* 143 (2002), pp. 5–19; António E. dos Reis, "Os instrumentos de medida de Pedro Nunes", *Oceanos* 49 (2002), pp. 20–34; and Alistair Kwan, "Vernier Scales and Other Early Devices for Precise Measurement", *American Journal of Physics* 79 (2011), pp. 368–373.

4 Pedro Nunes, *De Crepusculis liber unus*, Ludouicus Rodericus, Lisbon, 1542, fols. 17–18.

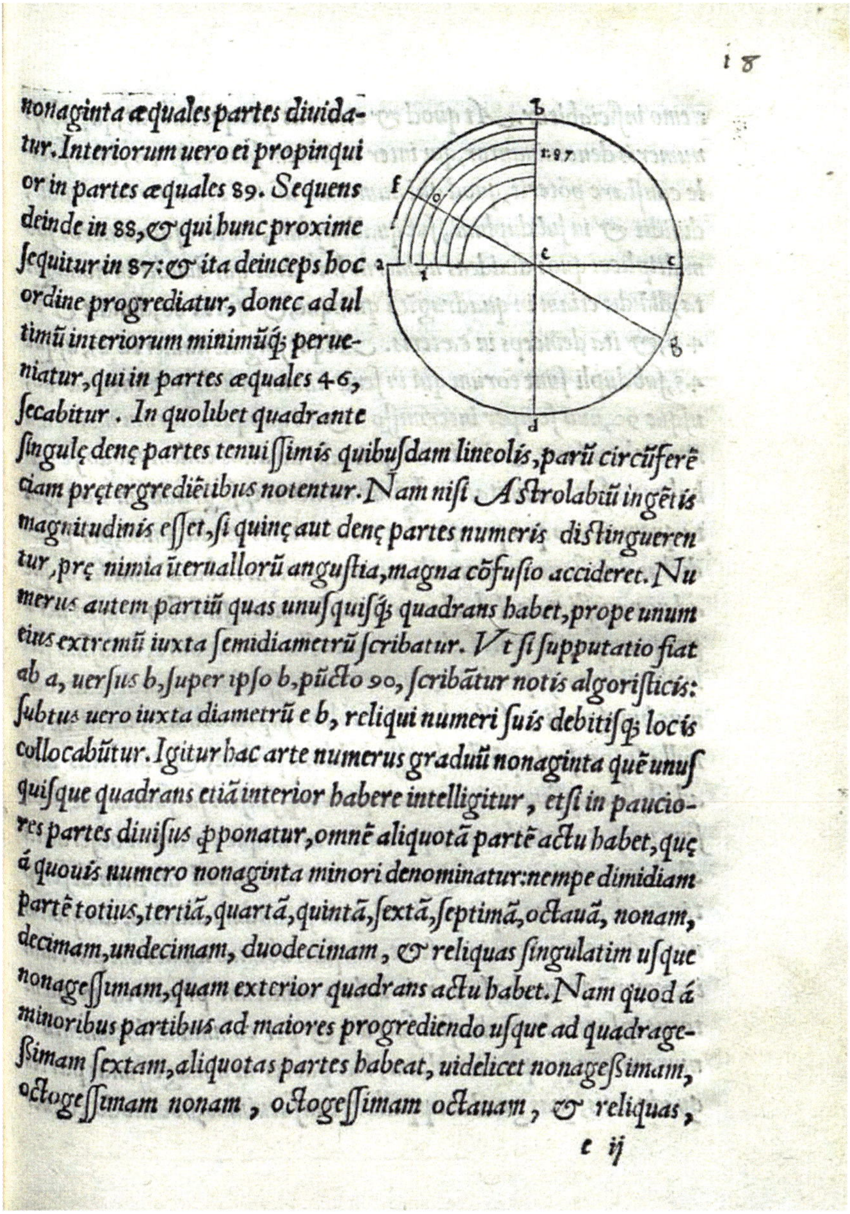

nonaginta æquales partes diuidatur. Interiorum uero ei propinquior in partes æquales 89. Sequens deinde in 88, & qui hunc proxime sequitur in 87: & ita deinceps hoc ordine progrediatur, donec ad ultimũ interiorum minimũq; perueniatur, qui in partes æquales 46, secabitur. In quolibet quadrante singulę denę partes tenuissimis quibusdam lineolis, parũ circũferẽciam prętergredientibus notentur. Nam nisi Astrolabiũ ingẽtis magnitudinis esset, si quinę aut denę partes numeris distinguerentur, prę nimia ĩteruallorũ angustia, magna cõfusio accideret. Numerus autem partiũ quas unusquisq; quadrans habet, prope unum eius extremũ iuxta semidiametrũ scribatur. Ut si supputatio fiat ab a, uersus b, super ipso b, pũcto 90, scribãtur notis algoristicis: subtus uero iuxta diametrũ e b, reliqui numeri suis debitisq; locis collocabũtur. Igitur hac arte numerus graduũ nonaginta quẽ unusquisque quadrans etiã interior habere intelligitur, etsi in pauciores partes diuisus ꝓponatur, omnẽ aliquotã partẽ actu habet, quę á quouis numero nonaginta minori denominatur: nempe dimidiam partẽ totius, tertiã, quartã, quintã, sextã, septimã, octauã, nonam, decimam, undecimam, duodecimam, & reliquas singulatim usque nonagessimam, quam exterior quadrans actu habet. Nam quod á minoribus partibus ad maiores progrediendo usque ad quadragessimam sextam, aliquotas partes habeat, uidelicet nonagessimam, octogessimam nonam, octogessimam octauam, & reliquas,

e ij

FIGURE 9.1 Pedro Nunes's book *De Crepusculis*, Lisbon, 1542, fol. 18r, where the author presents a diagram of the nonius scale
NATIONAL LIBRARY OF PORTUGAL, LISBON, BNP RES. 397 P

tropics as corresponding to the fraction 11/83.[5] This ingenious solution to a practical problem was created by a high level mathematician from an essentially theoretical perspective.

1 Tycho Tests the Nonius

In the last quarter of the sixteenth century, Tycho Brahe created a programme for the renovation of astronomy based on systematic observations. During his stay at Uraniborg, in a continuous quest for more accurate observations, he developed new instruments, increasing their size and introducing improved scales and sights.[6]

Tycho was familiar with Nunes's *De Crepusculis* and in his search for accuracy tried the nonius on four instruments. The nonius applied to the "small brass quadrant" and the "medium azimuthal brass quadrant" (*Fig. 9.2*) were shown explicitly in their respective engravings in the *Astronomiæ instauratæ mechanica*. However, the nonian solution was also tested in two other instruments: the "steel sextant" and the "large azimuthal semicircle".[7] In the *Astronomiæ instauratæ mechanica*, Tycho informed the reader that in the "small brass quadrant" (radius 39 cm) he had added a table to the reverse of the instrument in order to avoid the necessary calculations.[8] The same table was incorporated into the "medium azimuthal brass quadrant" (radius 58 cm). This was a more elaborate and significant instrument, since it was used in Tycho's observations of the comet of 1577, whereby measurement of the parallax, Tycho famously concluded that the phenomena was in the aethereal

5 Reis, "O Nónio" *op. cit.* (n. 3), p. 8. The cosmographer reveals his inspiration for the nonius in Pedro Nunes, *De arte atque ratione navigandi libri duo*, António de Mariz, Coimbra, 1573, p. 47. The reference in Ptolemy's work is in *Almagestū Cl. Ptolemei. Opus ingens ac nobile omnes celorū motus continens*, P. Liechtenstein, Venice, 1515, p. 9.

6 On Tycho's instruments and achievements there is a vast literature, see in particular: Victor E. Thoren, *The Lord of Uraniborg: A Biography of Tycho Brahe*, Cambridge University Press, Cambridge, 1990; Victor E. Thoren, "New Light on Tycho's Instruments", *Journal for the History of Astronomy* 4 (1973), pp. 25–45; and John R. Christianson, "The Legacy of Tycho Brahe", *Centaurus* 44 (2002), pp. 228–247.

7 Tycho Brahe, *Astronomiæ instauratæ mechanica*, Wandsbeck, 1598. I shall cite this book from the following English translation (hereafter referenced as *Mechanica*): Alena Hadravová, Petr Hadrava, and Jole R. Shackelford, *Instruments of the Renewed Astronomy*, Koniasch Latin Press, Prague, 1996.

8 Brahe, *Mechanica, op. cit.* (n. 7), p. 11. Other instruments where the nonius was tested, including comments on Pedro Nunes's invention, are in Brahe, *Mechanica*, pp. 17, 86. A fourth instrument where the nonius was used as an experiment – the "large azimuthal semicircle" – is pointed out in Thoren, "New Light", *op. cit.* (n. 6), p. 41.

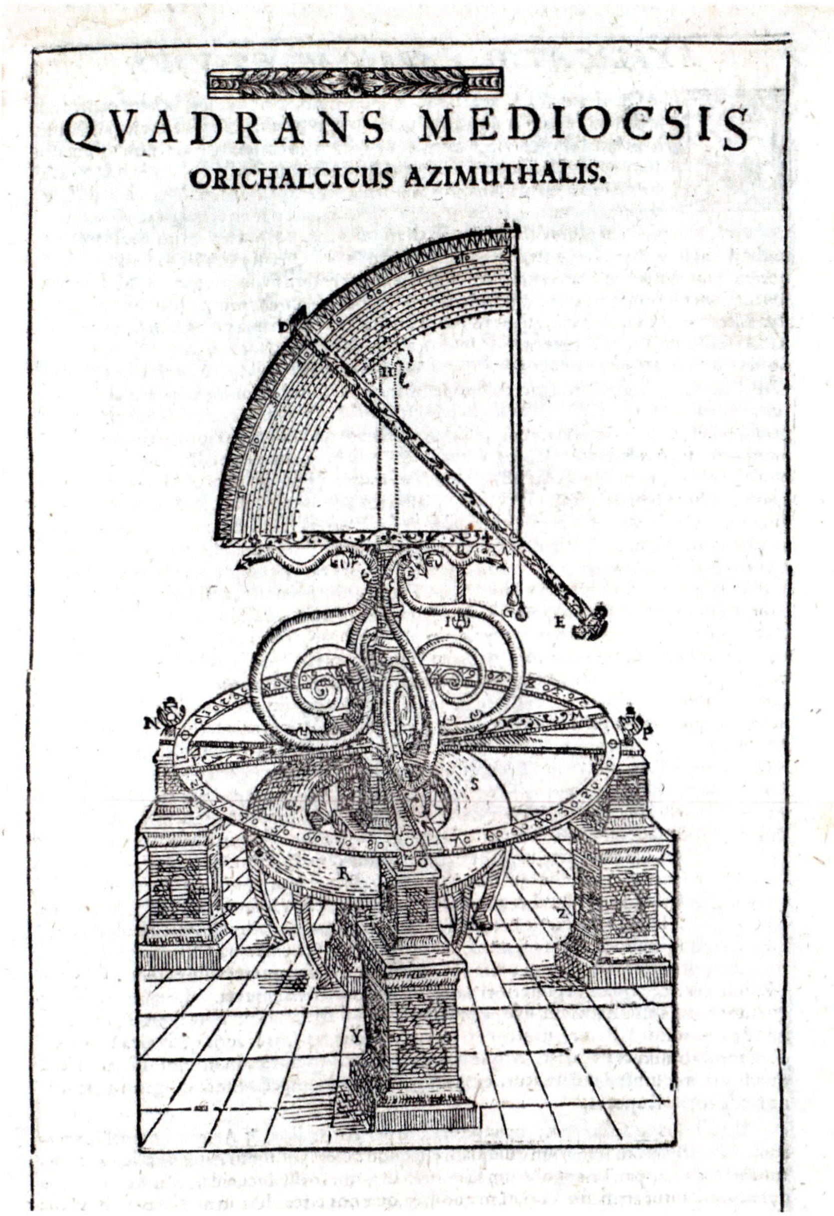

FIGURE 9.2 Engraving of the "medium azimuthal brass quadrant" with the nonius, in the second edition of Tycho Brahe's *Astronomiæ instauratæ mechanica*, Nuremberg, 1602
DIBNER LIBRARY OF THE HISTORY OF SCIENCE AND TECHNOLOGY, WASHINGTON, DC, Q QB85 .B7 1602

regions contrary to Aristotelian physics. The instrument allowed for measurements of the altitude (a vertical angle) and the azimuth (a horizontal angle) of any celestial body, and its nonius scale was manufactured with an additional number of divisions.

However, Tycho found the scales of the nonius difficult to make and read. In the end, he clearly advocated in the *Astronomiæ instauratæ mechanica* for the use of the diagonal scale instead of the nonius. As an authority in observational matters, Tycho's position helped to determine the reception of the nonius in the following centuries.

Nonetheless, several cosmographers and scholars concerned with navigation mentioned the nonius in their treatises. In Iberia, for instance, João Baptista Lavanha (*c.*1550–1624) in his *Tratado del Arte de Navegar* (1588), Francisco da Costa (1567–1604) in the work, *Arte de Navegar* (1596), and Andrés Garcia de Céspedes in the *Regimiento de Navegacion* (1606) discussed the nonius in some detail. For them, Nunes's proposal was useful and deserved to be adopted, raising the question of the effective use of the nonius by navigators, at sea, during the sixteenth and seventeenth centuries.

In England, George Waymouth (*c.*1585–*c.*1612) argued in his manuscript treatise "The Jewell of Artes", now preserved in two copies kept in the British Library and at Yale University Library, that the quadrant with the nonius was more accurate than one with a diagonal scale.[9] Indeed, he was very generous with the nonius, depicting it on six instruments in the treatise (even though he also depicted the diagonal scale on eight instruments).

Although several references to the nonius are found in the technical literature of the time, no evidence for its use at sea has survived. This might be either due to wooden quadrants with the nonius having not survived, or, as David Waters believed, a consequence that few navigators understood its principle and even fewer were prepared to go through the labour of doing the necessary calculations and converting each observation into degrees, minutes, and seconds of arc.[10]

9 George Waymouth, "The Jewell of Artes", 1604: London, British Library MS 19889, and New Haven, Yale University Library, Beinecke MS 565. See also Nicolàs de Hilster, *Navigation on Wood: Wooden Navigational Instruments 1590–1731*, published by Nicolàs de Hilster, Castricum, 2018, pp. 233–234.

10 David W. Waters, *The Art of Navigation in England in Elizabethan and Early Stuart Times*, Hollis and Carter, London, 1958, p. 304. Several sources indicate that wood quadrants were produced and used at sea. Ferdinand Magellan's expedition is an example, in spite of the fact that occurred too early to include the nonius, which was proposed two decades later. Six wood and fifteen wood-and-bronze quadrants are documented among the instruments used on Magellan's voyage; see Stefan Zweig, *Fernão de Magalhães: A Biografia*, Aletheia/Expresso, Lisboa, 2019, pp. 92–93.

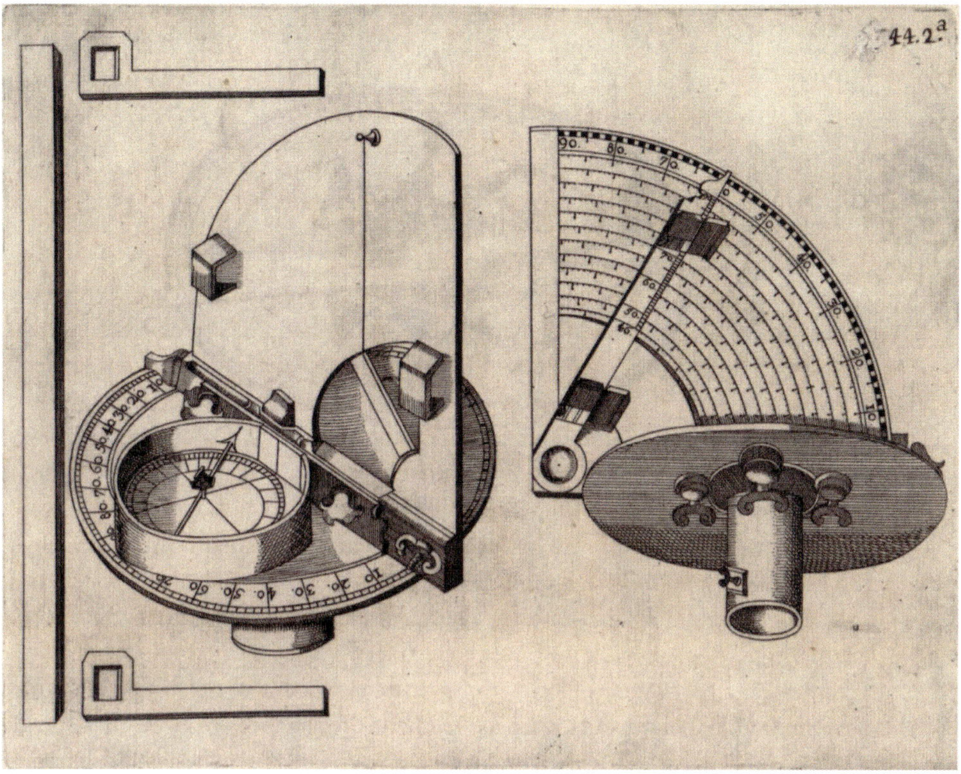

FIGURE 9.3 Azimuthal quadrant with a nonius scale, attributed to the English maker James Kynvyn, and produced for Robert Dudley, *c.*1595, as depicted in Dudley's *Dell'arcano del mare*, Florence, 1647
BIBLIOTECA RICCARDIANA, FLORENCE

It is also known that Jacob Kurz (1554–1594), the pro-chancellor of Emperor Rudolph II (1552–1612), and the prominent Jesuit mathematician Christopher Clavius proposed slightly different versions of the nonius. Kurz's work materialised into a quadrant made for Tycho Brahe that did not survive. The only known surviving early instrument, an azimuthal quadrant (*Fig. 9.3*) was made around 1595 by the English maker James Kynvyn (*fl.* 1570–1610) for Robert Dudley (1574–1649). Dudley was an English soldier, navigator, and cartographer who in 1605 left for Italy and became a geographer and shipbuilder in the service of the grand dukes of Tuscany in Florence. In 1646 and 1647, he published in three volumes the monumental *Dell'Arcano del Mare*, a summary of contemporary knowledge on navigation and possibly the first maritime atlas to encompass the whole world. In this work, he included an engraving of an azimuthal quadrant with the nonius scale. In order to increase legibility, it had only twelve graduated arcs and a scale on the alidade. This example shows that there was still space for improvement, especially in the domain of the use and legibility of the nonius. Dudley's azimuthal quadrant is now part of the Museo Galileo, linked to its Medici holdings.[11]

2 The Mystery of a Roman Jesuit Invention

Although sources abound for my first case, the nonius, they are very scarce for my second case. In 1720, Father Orazio Borgondio (1675–1740), chair of mathematics in the Roman College, proposed in the *Mémoires pour l'histoire des sciences et des beaux arts* (also known as the *Mémoires de trévoux*) a quadrant equipped with scales where, allegedly, minutes and seconds of arc could be directly read (*Fig. 9.4*).[12] Unfortunately, no diagram accompanied the text, making our efforts of interpretation very difficult (stressing once again the central importance of visual sources in the scientific and instrumental discourse of the period). Apparently, Borgondio's scheme used a second and smaller quadrant, a small telescope, a cylinder, and a full circle with a 360-degree graduated arc and a pointer, all mounted in the centre of gravity of the main quadrant. Screws were also employed, but without an illustration it is very difficult (virtually impossible?) to grasp how all the elements interacted.

However, the main focus of my analysis here is the reception and use of Borgondio's ingenious device. In that direction another source informs us that the same instrument, or another copy of the quadrant, produced by the same maker, Dominico Lusverg (*fl.* 1700–1720), was received by the Jesuits, in Lisbon, in 1724.[13] In the following years, the instrument was used for several observations in the Portuguese capital, by two Neapolitan Jesuits, Giovanni Battista Carbone (1694–1750) and Domenico Capacci (1694–1736). Carbone and Capacci had travelled to Portugal in 1722 in fulfilment of a royal request for Jesuit mathematicians to staff a geographical, cartographical, and astronomical mission in the extensive territory of Portuguese America (Brazil).[14]

---

11 Quadrant by James Kynvyn, *c.*1595, Museo Galileo, Florence (Italy), inv. no. 242, 3362; described by Gerard L'E. Turner, *Elizabethan Instrument Makers: The Origins of the London Trade in Precision Instrument Making*, Oxford University Press, Oxford, 2000, pp. 238–241.

12 Orazio Borgondio, "Nouvelle construction d'un quart astronomique dans lequel on a avec facilité et sureté les secondes de degrez", *Mémoires pour l'histoire des sciences et des beaux arts* (April 1720), pp. 764–766.

13 Giovanni Battista Carbone, *Observationes astronomicae habitae Ulissypone*, undated pamphlet containing astronomical observations from 1725 and 1726.

14 See Luís Tirapicos, "Astronomy and Diplomacy at the Court of King João v of Portugal", *Cahiers François Viète*, series 3, 3 (2017), 61–79.

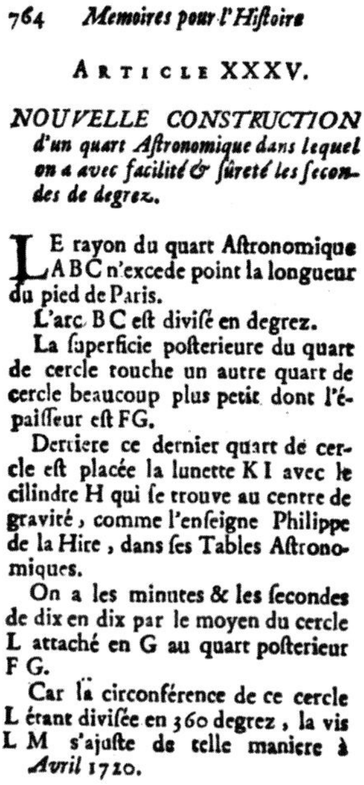

764 *Memoires pour l'Hiſtoire*

ARTICLE XXXV.

NOUVELLE CONSTRUCTION d'un quart Aſtronomique dans lequel on a avec facilité & ſûreté les ſecondes de degrez.

LE rayon du quart Aſtronomique ABC n'excede point la longueur du pied de Paris.

L'arc BC eſt diviſé en degrez.

La ſuperficie poſterieure du quart de cercle touche un autre quart de cercle beaucoup plus petit dont l'épaiſſeur eſt FG.

Derriere ce dernier quart de cercle eſt placée la lunette KI avec le cilindre H qui ſe trouve au centre de gravité, comme l'enſeigne Philippe de la Hire, dans ſes Tables Aſtronomiques.

On a les minutes & les ſecondes de dix en dix par le moyen du cercle L attaché en G au quart poſterieur FG.

Car la circonférence de ce cercle L étant diviſée en 360 degrez, la vis LM s'ajuſte de telle maniere à

Avril 1720.

FIGURE 9.4
First page of Orazio Borgondio's paper describing his invention of "a new construction of an astronomical quadrant in which we have easily and with certainty the arc second", *Mémoires pour l'histoire des sciences et des beaux arts* (April 1720), pp. 764–766
GOOGLE BOOKS/BAVARIAN STATE LIBRARY, MUNICH

According to Ugo Baldini, Orazio Borgondio was the teacher who partially restored the vigour of mathematical teaching at the Roman College in the eighteenth century, following its technical decline at the end of the previous century.[15] Borgondio, who held the chair of mathematics between 1712 and 1740, is mainly known as the teacher and predecessor of the notable natural philosopher Ruggero Boscovich (1711–1787).[16] But Borgondio had many other disciples, and Giovanni Battista Carbone and Domenico Capacci were among them. In two letters that Carbone wrote to the Roman College in 1723 and 1724, addressed to Filippo Buonanni (1638–1725), he sent a warm hello to "Fr. Borgondio my teacher", and "A thousand greetings of thanks and reverence to Fr. Borgondio my godfather, master and friend".[17]

15 Ugo Baldini, "The Academy of Mathematics of the Collegio Romano from 1553 to 1612", in M. Feingold (ed.), *Jesuit Science and the Republic of Letters*, MIT Press, Cambridge, MA, 2003, pp. 47–98: 53.

16 On Boscovich see Ugo Baldini, "Boscovich e la tradizione gesuitica in filosofia naturale: continuità e cambiamento", *Nuncius* 7 (1992), pp. 3–68.

17 "*Il P. Borgondio mio maestro*" and "*Mille saluti ringraziamente e riverenze al P. Borgondio al mio Compadre, Padroni ed Amici*". Giovanni Battista Carbone, 14 June 1723, National Central Library Rome, A. 194/531, and Giovanni Battista Carbone, 31 July 1724, National Central Library Rome, A. 194/532. From 1698 Filipo Buonanni was the Jesuit naturalist responsible for the reactivation of the Kircherian Museum, and the author of its new catalogue produced in 1709.

Borgondio was above all an empiricist who saw experience and observation as the foundation of true physics. For him, data generated by celestial observations was the decisive validation for all astronomical calculations and theories. Starting in 1713, Borgondio authored many mathematical pamphlets – although frequently published under the generic signature "ex. PP. Societatis Iesu" – in which he dealt with matters ranging from cartography, mechanics, geometry, and geography, to the computation of calendars. In the 1720s and 1730s he was also an active observer of celestial phenomena such as eclipses or planetary transits.[18] His close relationship with astronomical instruments – and his knowledge of the subject – was evident when, after being appointed superintendent of the Kircherian Museum in 1725, he created a new room that was especially dedicated to them. So, it is no surprise that Borgondio in 1720 would propose and publish in the *Mémoires des Trévoux* a new device to improve the accuracy of angular scales, allegedly allowing readings of the order of a second of arc.[19] It was evidently an important effort to improve the quality of astronomical instruments for angular measurements, which a few years later, would be applied to one of the quadrants ordered for Carbone and Capacci in the court of Lisbon.[20] Thus, everything suggests that the stay of the two Neapolitan Jesuits at the Roman College, in the summer of 1722, learning new observational techniques with Borgondio would have provided a background that would mark their life and scientific activities in the years that followed.

No other references are known to the adoption and use of Borgandio's invention. Nevertheless, several decades later, in Prague, another Jesuit mathematician Joannes Klein (1685–1762) tried what seems to be a similar solution in a quadrant that remained unfinished in 1762. This instrument in now preserved in the National Technical Museum (NTM no. 17 135), and the alidade of the quadrant displays dials for direct reading of minutes and seconds of arc, according to Zdislav Šíma "something that was completely unattainable by the simple mechanical gears used".[21]

18 Brendan Dooley, "The *Storia Letteraria d'Italia* and the Rehabilitation of Jesuit Science", in Feingold, *op. cit.* (n. 15), pp. 433–473: 437, 464.

19 Borgondio, *op. cit.* (n. 12). Borgondio is completely ignored in standard histories of astronomical instrumentation showing that several Jesuit contributions are still to be properly investigated and evaluated. See, for instance, J. A. Bennett, *The Divided Circle: A History of Instruments for Astronomy, Navigation and Surveying*, Phaidon/Christie's, Oxford, 1987.

20 Carbone, *op. cit.* (n. 13).

21 Zdislav Šíma, "Prague Sextants of Tycho Brahe", *Annals of Science* 50 (1993), pp. 445–453: 453.

3 Concluding Remarks

These two contrasting cases of failed schemes for reading angular scales show us that to be widely adopted an instrument must not only be accurate and precise but its manufacture and use should be as simple and friendly as possible. In fact, this is particularly clear in the case of the nonius and the later wide adoption of the vernier scale, a much simpler and more legible alternative. And if in the eighteenth century several nautical and astronomical instruments were still furnished with a "nonius" this was more a result of the fame of the nonian invention than of its actual use. Navigational and surveying instruments were often said to have a nonius, when what they had, in fact, was a vernier scale. The word "nonius" was misapplied to the vernier. The latter was, in effect, a different concept; but for many users and makers somehow related with, or originated in, Pedro Nunes's ingenious proposal.

As for Borgandio's device, it remains a mystery why it was so quickly forgotten and abandoned, but it could have been too complicated to work well mechanically, as happened with a similar solution tested in a quadrant of Joannes Klein SJ, in Prague. Perhaps the lack of references outside of Jesuit circles may account for the limited circulation of the instrument, essentially restricted to the network of the Society's mathematicians and astronomers. This lack of marketing and possibly other technical difficulties may be among the reasons for the apparent obscurity and abandonment of the invention after the 1720s.

Acknowledgements

A warm thanks is due to Nicolàs de Hilster for all his kindness and generosity sharing documents and his knowledge of George Waymouth. This research was funded by the Portuguese *Fundação para a Ciência e a Tecnologia* (FCT CEECIND 02607 2017).

CHAPTER 10

Thomas Young's Eriometer: Wrong Time, Wrong Inventor, Wrong Users?

Michelle Mercier

Thomas Young (1773–1829) is widely known among physicists today for his double-slit experiment of 1807. However, the British natural philosopher and physician carried out many other optical experiments and inventions during the course of his extensive work on the wave nature of light and the principle of interference. One instrument by Young was a practical application of his optical investigations on diffraction yet remains largely unknown: the eriometer.

The instrument is based on the following idea: Light enters the eye of the user after passing through a central hole in the middle of a perforated plate and then through a sample of fibres or particles, which is affixed to a comb or card held near the eye. The light is diffracted by the fibres or particles, and creates a diffraction phenomenon in the eye. From this diffraction pattern, the user can infer the size of the fibres or particles.

Young claimed to be able to analyse the diameter of small fibres and particles with this instrument, for which he offered several applications. Believing that the instrument was well suited to estimate the quality of wool fibres, Young aptly named the instrument the "eriometer", after the Greek word for wool (*erion*). He also advocated for the use of his instrument in medical contexts, arguing that – among other potential applications – it could help to distinguish between mucus and pus, since the dimensional homogeneity (size similarity) of pus "globules" would produce a diffraction pattern for these cells, whereas mucus, containing no "globules", would not. The instrument could be constructed with simple materials, such as a paper card and a piece of tape. Young's eriometer was the first technical application of the diffraction principle to use diffraction effects to determine the size of physical objects.[1]

1 Jens Weber, "Thomas Youngs Eriometer: Das erste beugungsoptische Meßinstrument zur Bestimmung von Faser- und Partikeldurchmessern", unpublished diploma thesis, Carl von Ossietzky Universität Oldenburg, 2001. Also see Peter Heering and Jens Weber, "Größenbestimmung durch Beugung", *Physik in unserer Zeit* 33 (2002), p. 92.

 | DOI:10.1163/9789004689107_012

Although practical experiments showed that an experienced user could obtain precise enough results using Young's measuring method, the eriometer can be described as a failure in the sense that, in the early 1800s, it failed to become firmly established in the English medical, agricultural, or textile industries. The factors that led his contemporaries in various communities to reject the eriometer are probably complex. With that in mind, this study does not attempt to offer a monocausal explanation for this failure.

In London, I was able to identify two surviving eriometers by Young. This paper analyses both objects, examines Young's eriometer from different perspectives, and speculates about the possible factors for the instrument's contemporary rejection in England.

1 A Gift for Sir Joseph Banks (1810)

One eriometer is part of the collection of the British Library and was not explicitly catalogued (*Fig. 10.1*). Young enclosed it in a letter he sent on 10 September 1810 to Joseph Banks (1743–1820), president of the Royal Society at the time.[2]

The object is a paper card, blackened on one side and marked "Dr Young; 18, Welbeck Street". On the other side is the handwritten headline "The Agricultural Micrometer", a text, and a sketch of two concentric circles. The card has been pierced through in several places along the circumferences of the circles, and once at the common centre of both. Through a small vertical slit in the card, a long tape has been passed. A second card is attached to one end of the tape and has a little cut-out window through which some fibres can be seen. The fibres are held in place by a folded piece of paper inserted into the card.[3] As understood from the letter, this second card was meant to hold the

2 Thomas Young to Joseph Banks, 10 September 1810: London, British Library, Letters addressed to Sir Joseph Banks, Add. MS. 33982, fols. 9–10*.

3 The handwritten text was supposed to be an instruction for the use of the instrument, as Young stated in his letter to Banks, noting that the instructions "could be engraved and printed on the card". It can also be taken from the letter that the central hole should have been 1/20 inch in diameter. The diameter of the inner circle 11/20 inches and the outer one a diameter of one inch. The diameters of the apparently smaller holes on the concentric circles have not been explicitly stated by Young, see Young, *op. cit.* (n. 2), fol. 9. Young also stated in this letter that the fibres of wool he prepared for Banks were fibres of the Vigonia wool, "which exhibits the colours in great perfection". Whether the fibres in the surviving instrument are in fact still those fibres cannot be said with certainty, since Banks had "tried to use" the instrument, as can be deduced from a later letter by Young to Banks, dated on

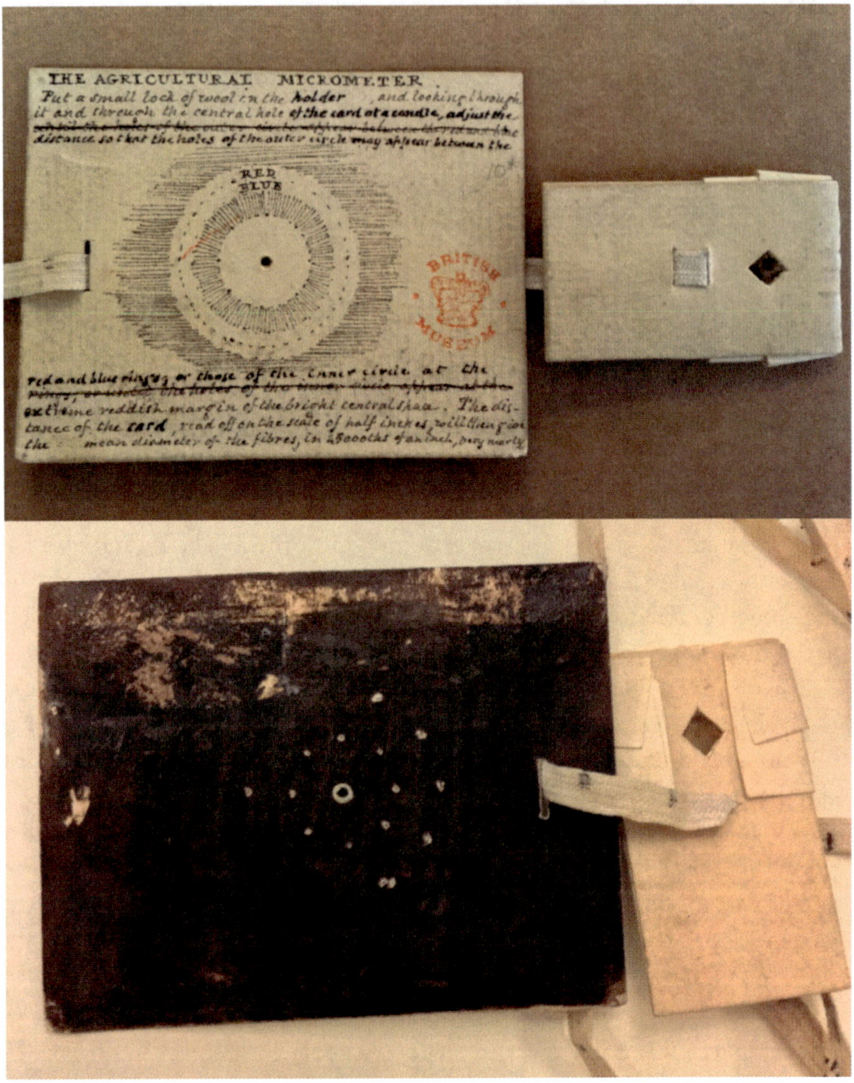

FIGURE 10.1 Thomas Young's eriometer. Part of a letter from Young to Sir Joseph Banks dated 10 September 1810

6 October 1810. A comparison of the dimensions with Young's noted dimensions in his first letter to Banks showed that only the diameter of the central hole was slightly larger than indicated – the diameter is about 1.5/20 inches. This discrepancy can be traced back to Young's recommendation to increase the diameter slightly for a better observation in the later letter

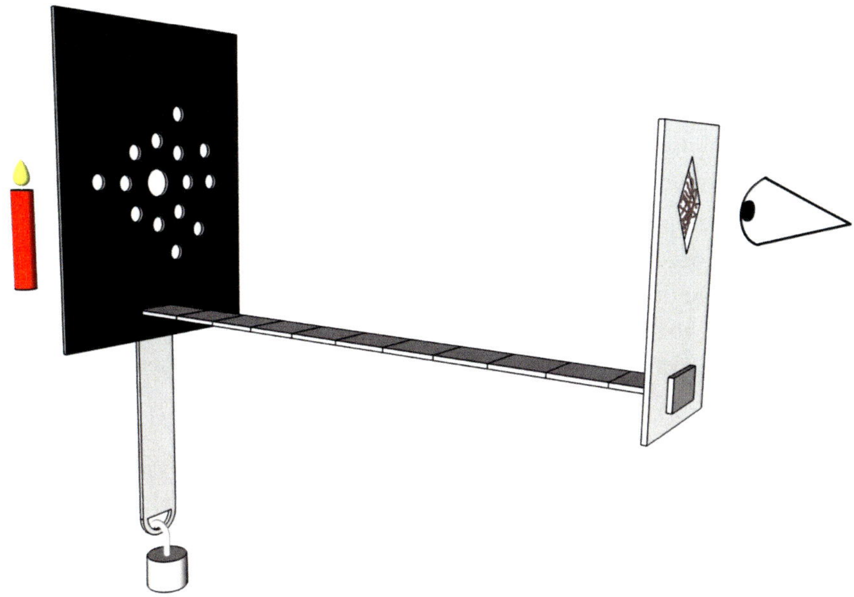

FIGURE 10.2 A schematic sketch of the experimental setup
ILLUSTRATION BY PHIL HANISCH, 2020, REPRODUCED BY PERMISSION

wool-fibre samples. Fixed to the other end of the tape is a ring, which cannot be seen in *Figure 10.1*. As the letter makes clear, a small weight can be attached to the ring to tighten the tape.[4] A schematic sketch of the arrangement can be seen in *Figure 10.2*. The tape itself has 70 marks, spaced out at half-inch intervals. This was obviously a measuring tape.

2 A Previously Unknown Eriometer by Young (1810–1813)

The archive of the Royal Institution also lists an instrument, catalogued as "the Eriometer", which has been in its collection since about 1960.[5] From a closer examination of the instrument's function, materiality, and construction, it can be deduced that this instrument is also an eriometer by Thomas Young, with a construction that very likely dates back to 1810–1813.

---

to Banks, see Thomas Young to Joseph Banks, 6 October 1810: London, British Library, Letters addressed to Sir Joseph Banks, Add. MS. 33982, fol. 7.

4 Young to Banks, *op. cit.* (n. 2), fol. 9.

5 "The 'Eriometer'": London, Royal Institution, RIAC 00184. I am currently not able to throw any light on how the instrument arrived at the Royal Institution. The archive number only suggests that it came into their collection around 1960.

Although this object does not have (or no longer has) a tape, the perforated paper card has a slit similar to that on the eriometer Young made for Banks in 1810. The card is also blackened on one side and perforated along the perimeter and centre of two circles. In contrast to the Banks eriometer, the circles on this perforated card are not concentric. Thus, the perforations along the perimeter and at the centre point are separate for each circle. However, in this regard, the second eriometer fits with Young's only public description of his instrument in 1813, which also corresponds to the device kept at the Royal Institution in terms of the diameters of the circles bearing the perforated holes, and the diameters of the two central holes.[6]

The object also includes four sample holders made of paper card with a variety of different wool fibres. In addition to the fibre samples they hold, the sample cards are inscribed with measurement notations. Most of them can be found in Young's 1813 publication "4. Substances measured by the Eriometer".[7]

The sample cards are slotted in a different manner than that found on Banks's eriometer. The slots enabled the sample cards to be placed and moved on a slider. Both Young's letter to Banks of 1810 and his public description of the eriometer in 1813 state that Young sometimes used a small rod of wood (1810) or a slider construction (1813) instead of the tape. These were not the only variations in Young's designs for the eriometer. Instead of a perforated paper card, Young at times used a pierced brass plate (1813), and instead of a sample-holder (as shown in *Figure 10.1*), he sometimes placed several pins at the end of a wooden rod (1810) or fixed the fibres to be examined "on some wires, which are carried by a slider" (1813).[8] These modifications did not make the eriometer a different instrument for Young, although he attributed greater accuracy to the version of the instrument that was not just paper and tape and thus less portable.[9]

---

6 Thomas Young, "Remarks on the Measurement of Minute Particles, especially those of the Blood and of Pus", in Thomas Young, *An Introduction to Medical Literature, Including a System of Practical Nosology: Intended as a Guide to Students, and an Assistant to Practitioners*, B. R. Howlett, London, 1813, pp. 545–560. Reprinted also in *Annals of Philosophy* 2 (1814), pp. 115–122 and 190–195.

7 *Ibid.*, p. 552.

8 For noted details about Young's eriometer from 1810, see Young to Banks, *op. cit.* (n. 2), fol. 9; for remarks and quotations in 1813, see Young, "Remarks", *op. cit.* (n. 6), p. 549.

9 Young, "Remarks", *op. cit.* (n. 6), p. 549. In 1810, Young had only noted an easier transport of the paper and tape device; see Young to Banks, *op. cit.* (n. 2), fol. 9.

3 A Measuring Instrument – for What?

Regarding the function of his eriometer, Young wrote in his 1810 letter to Banks that it could be used to determine the fibre diameter of various types of wool. He noted that farmers, however, would be more interested in fibre uniformity – which was (and still is) a quality marker for wool – and that the instrument he enclosed should also make it easy for them to examine this by enabling them to estimate the extent to which a certain number of fibres would have nearly the same diameter.

His eriometer was also intended to have a practical function in the medical field,[10] as Young makes evident in his description of the instrument in 1813, published in his *Medical Literature* under the title "Remarks on the measurement of minute particles, especially those of the blood and pus".[11] His eriometer, Young noted (referring to it as such in this publication), allowed him to determine the size not only of wool fibres, but also of particles like red blood cells and capillary arteries. Hence, he commented, it should also allow medical scientists to examine bodily fluids (blood and pus, for example) to see whether they contained particles of approximately the same size and, by means of the diffraction rings produced, to determine the size of the particles. Since it should not be possible to create diffraction rings with mucus, he argued, the instrument should enable investigators to differentiate between mucus and pus. Given that pus is an indication of infection, the instrument could also have a diagnostic function in medical practice and, further, could give physicians the advantage of being able to identify the diameter of cells accompanying diseases.[12]

4 The Measuring Principle from Today's Point of View

From a modern viewpoint, we can easily understand the measuring principle of Young's eriometer (*Fig. 10.2*). The perforated plate would have been held in front of a candle flame in such a way that the candlelight would enter the user's

10 He had made first hints in this regard already in his letter to Banks from 1810: "(I have sometimes thought of employing the instrument as a nosological test of the state of the blood, of pus, and of other animal fluids)". Young to Banks, op. cit. (n. 2), fol. 9.

11 Young,"Remarks", *op. cit.* (n. 6).

12 Today, the value determined, MCV (mean corpuscular or cell volume) plays an important role in the diagnosis of various types of disease. The MCV value in a sample is compared with the normal value. An increased MCV value can be caused by a deficiency of vitamin B-12 or folic acid, for example. A low MCV level can often be caused by iron deficiency.

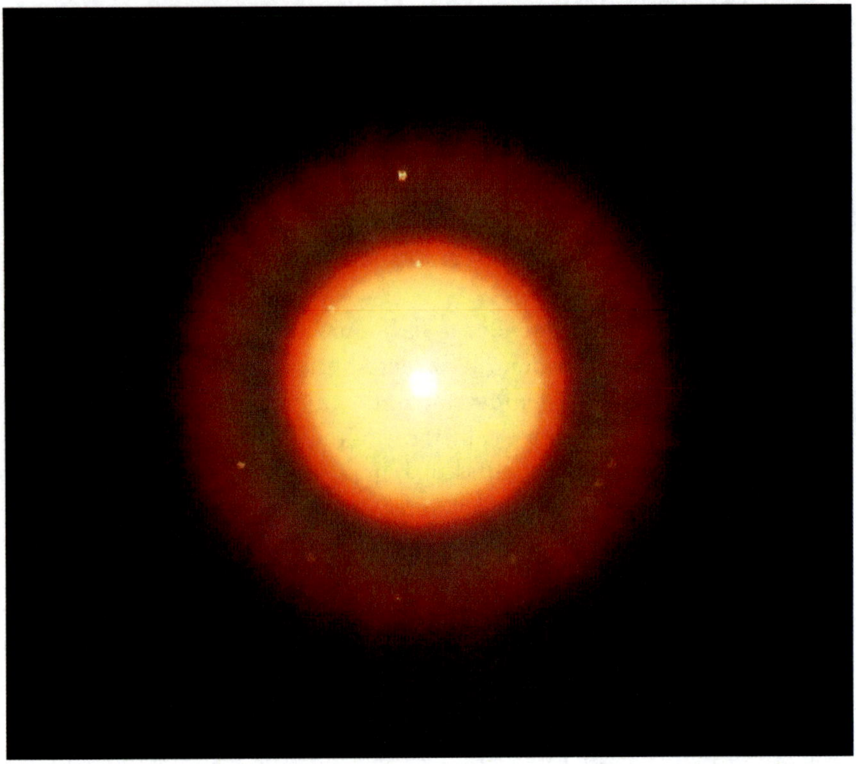

FIGURE 10.3 A diffraction phenomenon created by wool fibres examined with a replica of an eriometer
PHOTO BY PHIL HANISCH, 2020, REPRODUCED BY PERMISSION

eye through the plate's central hole and the sample. The transmitted light was diffracted by the fibres or particles and varied according to the wavelength; the respective colour components of the diffraction patterns would overlay each other in the eye. Then coloured diffraction rings would be observed (*Fig. 10.3*). The smaller the diameter of the respective fibres, the larger the diffraction angle would be and, thus, the larger the observable diffraction ring. Therefore, the size of the diffraction ring would be a function of the diameter of the sample elements.

To determine the size of wool fibres or particles, the distance between the perforated plate and the sample would be varied until one of the perforated circles coincided with a predetermined colour of the generated diffraction ring. Since the size of the diffraction ring – and thus the distance between the perforated plate and the sample – depended on the size of the respective sample elements, the size of the particles or fibres examined could be deduced by measuring the distance and comparing it to results from a sample of known size. For an absolute determination, however, the instrument must first be calibrated with a sample of known size to determine a conversion factor. A condition for the observation of a well-defined (distinct) and thus measurable diffraction phenomenon was the relative homogeneity in size of the sample elements. If the sizes of the sampled particles or fibres differed too much from each other, no measurable diffraction pattern could be observed. This is an advantage of the device: the phenomenon of diffraction rings could not be observed in very inhomogeneous samples, since this requires a reasonably homogeneous sample from which the size of particles or fibres most frequently occurring in the sample can be determined. The instrument therefore no longer worked if a measurement would have been meaningless anyway.

5 Wrong Time?

In early nineteenth century England, the underlying theory for Young's instrument was far from clear. The assumption that light was wave-like in nature was not widely accepted, nor would Young's attempt to establish a wave theory of light during the first decade of the century change this. This was still the case in 1813, when Young first publicly described his eriometer. Young himself, in fact, did not fully understand the theory on which he had based the instrument's measurement principle and construction: "The theory, which suggested to me the construction of the eriometer … depend[s] on circumstances not completely understood: at present therefore I shall only employ, for the determination of the true value [of the size of particles or fibres] … an experimental comparison."[13]

Between 1799 and 1804, Young published a series of papers on the wave theory of light and the principle of optical interference. However, his scientific work on physical optics received little public attention, apart from the public criticism from the Scottish-born Henry Peter Brougham (1778–1868). This lack of public notice can be seen as a form of rejection, and indeed was seen as

13 Young, "Remarks", *op. cit.* (n. 6), p. 551. For example, a theoretical description of the measuring principle or an explanation of the phenomena created thereby would have required an adequate conceptual understanding of diffraction by a narrow obstacle, which presupposes a theory of single-slit diffraction and Babinet's principle (first formulated in 1837).

such by Young, as revealed in a letter from Humphry Davy to Davies Gilbert on 26 October 1802: "Have you seen the theory of my colleague, Dr. Young, on the undulations of an ethereal medium as the cause of light? It is not likely to be a popular hypothesis after what has been said by Newton concerning it. He would be very much flattered if you could offer any observations upon it, whether for or against it."[14]

However, the failure of Young's instrument cannot only be explained by way of reference to British scientists' widespread belief in a particle nature of light, Young's inability to set forth a comprehensive theory, or the rejection of his optical works. To make use of the instrument for the purpose of practical measurements, there was no need to understand and apply the theory of diffraction underlying its measuring principle, nor did an instrument always require a transparent theory in order to achieve authoritative status.

Young saw an argument for the usability of his instrument in its simple construction and accuracy as a measuring device; as he told Banks in his first letter: "I should imagine it to be sufficiently accurate for all practical purposes, and the great facility and cheapness of its construction may perhaps render it useful to a class of persons, who would object to the expense of providing themselves with a mere complicated apparatus."[15]

Experimental studies of Phil Hanisch (2020) with a reconstruction of Young's eriometer for Banks from 1810 have shown that his targeted circle of users, especially farmers, could be seen as a decisive factor in the rejection of the instrument – albeit for different reasons.[16]

6 Wrong Users?

First of all, it can be stated that a replica of Young's eriometer from 1810 worked (*Fig. 10.4*): In the end it was possible for Hanisch to measure the size of wool

14 Humphry Davy, quoted in Henry Bence Jones, *The Royal Institution: Its Founder and Its First Professors*, Longmans, Green and Co., London, 1871, p. 330.

15 Young to Banks, *op. cit.* (n. 2), fol. 9. At that time, the measurement of the diameter of small fibres or particles would otherwise have meant the purchase and use of a suitable microscope.

16 Phil Hanisch, "Eine Analyse vom Eriometer nach Young mit der Replikationsmethode", unpublished bachelor thesis, Europa-Universität Flensburg, 2020. Also see Jens Weber, "Thomas Youngs Eriometer: Das erste beugungsoptische Meßinstrument zur Bestimmung von Faser- und Partikeldurchmessern", unpublished diploma thesis, Carl von Ossietzky Universität Oldenburg, 2001.

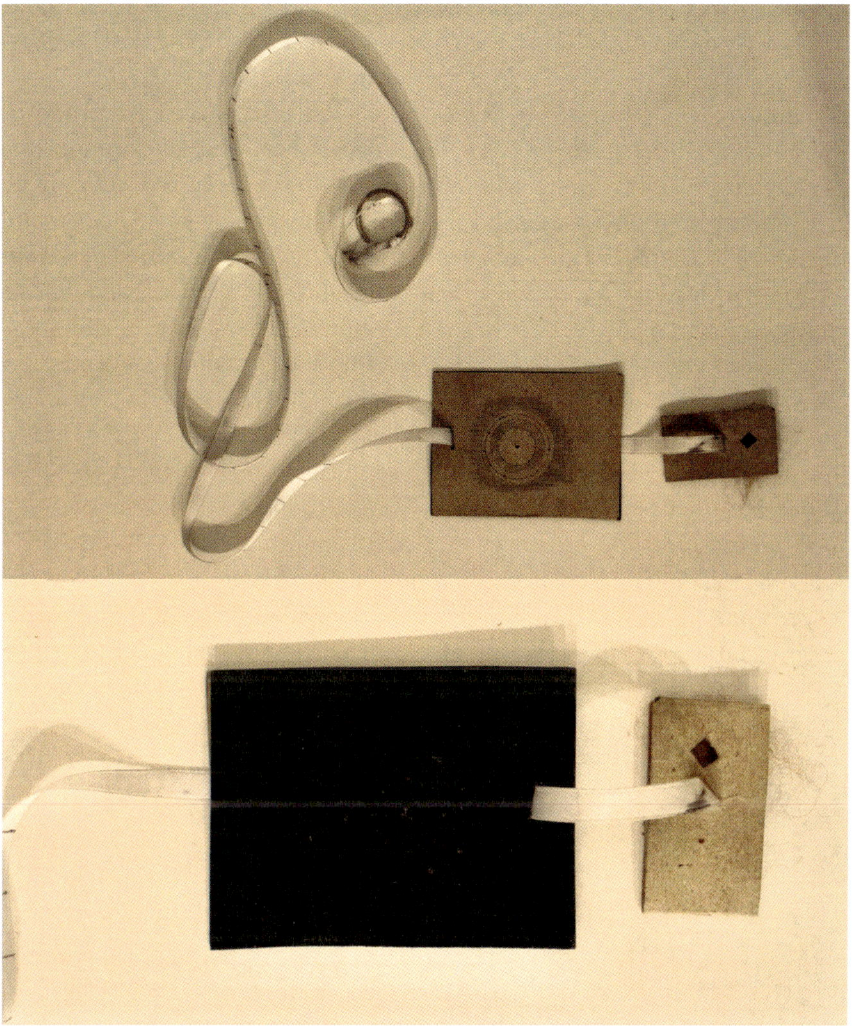

FIGURE 10.4 Replica of Young's eriometer for Banks, 1810, built by Phil Hanisch
PHOTO BY PHIL HANISCH, 2020, REPRODUCED BY PERMISSION

fibres with a high degree of accuracy. However, the path to this result was a long one. The fact that the experimenter could only observe a diffraction pattern but could not directly examine the sample (as, for example, one could with a microscope) made its practical use considerably more difficult at first. Not only did the handling of the instrument have to be learned, but also what should be seen and measured. In this respect, the use of a reference sample of very high quality wool fibres at the start of experimentation would be very helpful, giving a feeling for what could be seen and a practice in estimating the quality of the fibres. It is understandable that Young would have prepared samples of wool fibres for Banks to use when testing the eriometer.

One problem in learning to use the eriometer was that the extent to which the perceived diffraction ring was well-defined (or not) depended on more than the uniformity of the diameters of the fibres. If the colours observed were not very intense or distinct, this could also be due to the arrangement and amount of fibres in a sample or to the light conditions in the room. Experimental investigations of Hanisch have shown that, although the black background of the perforated card made it easier to see the respective colours of the diffraction ring, other light sources in the room – apart from the candle – could make the observation difficult to almost impossible. In addition, it was difficult to hold the eriometer in such a way that the light of the candle always passed through the central hole and sample into the eye during a measurement (*Fig. 10.5*). When the slightest breeze caused the candle to flicker, the diffraction rings were no longer visible. Moreover, they could not be observed if the eye of the examiner was not relaxed. In short, good laboratory conditions and an experienced eye were necessary to see diffraction rings at all. Before these diffraction rings could be used for a measurement, they had to be stabilised. To achieve

FIGURE 10.5 Experimental measurement of the diameter of small fibres by Phil Hanisch using his replica of Young's 1810 eriometer
PHOTO REPRODUCED BY PERMISSION

this, the user had to acquire specific knowledge regarding the handling of the instrument, preparation of the samples and adjustment of the eye.

For Young, who had already investigated diffraction phenomena for years and had studied that caused by fibres, these aspects did not pose any challenges for him around 1810.[17] But what about the circle of users to whom Young had addressed his instrument? For them, a substantial amount of competencies were to be acquired, which either would require a lot of time or a steep learning curve. Examining diffraction rings were not part of the common practice of farmers (also not in the medical context), which would have made it difficult for them to immediately use the instrument for measurement purposes. To what extent did typical farmers in about 1810 have a room that was absolutely sheltered from the wind, could be always darkened enough for an observation, and had a straight burning candle flame, is also an open question. Did they even want to measure with an instrument that was made of paper and tape and required such sensitive handling? Young himself blamed the latter for the rejection of his eriometer within the fields of agricultural or textile economics: "Dr Young has made this appearance the foundation of a mode of measuring the fineness of wool, which he has recommended for agricultural purposes, though it seems hitherto to have been found much too delicate to be employed by 'the hard hands of peasants,' with any advantage."[18]

In the medical field, Young did not directly comment on the failure of his eriometer. His *Medical Literature* was "intended as a guide to students, and an assistant to practitioners".[19] At the start of the nineteenth century, the formal distinction between physicians, surgeons, and apothecaries was still being made in England. Whereas physicians were responsible for diagnoses, prognoses, recommendations, and the prescription of drugs and had a university education, the work of a surgeon was compared to a craft, although it had been an independent profession since 1745. Surgeons were responsible for external complaints, such as skin diseases, wounds, and injuries, and they performed simple operations. While physicians worked with their heads, surgeons worked with their hands. At the beginning of the nineteenth century, apothecaries were also allowed to make recommendations, even if they were only free of charge, in addition to selling medicines. There were also the so-called practitioners who had gained experience or professional training as

17 Thomas Young, "An account of some cases of the production of colours, not hitherto described", *Philosophical Transactions of the Royal Society* 92 (1802), pp. 387–397.

18 Thomas Young, "Chromatics", in M. Napier (ed.), *Supplement to the fourth, fifth, and sixth editions of the Encyclopaedia Britannica*, Edinburgh, 1824, 6 vols.: v. 3, pp. 141–163: 150.

19 Young, *An Introduction to Medical Literature, op. cit.* (n. 6), title page.

surgeons, apothecaries, or both, but practised as a kind of general practitioner in the modern sense of the word. These practitioners often lacked the anatomical or scientific training of physicians, but still worked as apothecaries, surgeons, and physicians rolled into one. In the countryside, practitioners were often patients' only recourse.[20] Young called himself a practical physician and addressed his eriometer to practitioners in the title of his work. It is safe to assume, therefore, that he addressed not only those who had formal training as physicians and primarily worked with the head, but especially also practitioners who were educated or had previously worked as surgeons or apothecaries.

Young, however, was a physician with little practical experience. Did his eriometer simply target the wrong users' group – practitioners – because it had a diagnostic function, an indirect measuring principle, and a need to learn a specialised way of seeing and handling the instrument before using it? If we examine not only the user's perspective, but also the person of Thomas Young himself – the inventor – and his reputation as a practical physician, a different picture emerges.

7 Wrong (Status of) Inventor?

Thomas Young was born on 13 June 1773 in Milverton, a town in southwest England. In 1793, when Young was only twenty years old, a paper by him was first read at a meeting of the Royal Society and published in the *Philosophical Transactions*. In this publication, he first explained the accommodation of the eye the way we would describe it today.[21] He was elected a Fellow of the Royal Society one year later. He studied medicine in London, Edinburgh, and Göttingen, where he received his degree as a Doctor of Physic, Surgery, and Midwifery. After enrolling at Emmanuel College in Cambridge for two more years and thus completing his desired medical education, he was able to settle down in London in 1799 as a physician, opening his own practice there.

In addition to his work as a practical physician, Young regularly attended meetings of the Royal Society and devoted himself more and more to questions of natural philosophy, especially physical optics. At a meeting of the Royal Society, he also met the founder of the Royal Institution, Count Rumford

20 See Roy Porter, *Disease, Medicine and Society in England, 1550–1860*, 2nd. edn., Cambridge University Press, Cambridge, 1995; William F. Bynum and Roy Porter (eds.), *Companion Encyclopedia of the History of Medicine*, 2 vols., Routledge, London, 1993.

21 Thomas Young, "Observations on vision", *Philosophical Transactions of the Royal Society* 83 (1793), pp. 169–181.

(Benjamin Thompson). It was Rumford who recommended Young for the Chair of Natural Philosophy of the Royal Institution in 1801, a position he held alongside Humphry Davy, who was Professor of Chemistry at the time. After two years, Young retired from the position, which included giving lectures and publishing a journal together with Davy. He justified this decision to the Royal Institution by saying that this position was "so incompatible with the pursuits of a practical physician".[22] Shortly afterwards, however, he accepted a position as foreign secretary of the Royal Society. He was elected as a fellow of the Royal College of Physicians in 1809 and worked as a physician in a hospital from 1811 onwards. In about 1810, Young also gave another lecture series on physiology, chemistry, and nosology (the classification of diseases), but now at the Middlesex Hospital in London. In the public sphere, Young thus moved between two disciplines: science or natural philosophy and medicine.

For practitioners, the rejection of Young's work on physical optics did not necessarily mean a rejection of Young as a practical physician and his eriometer. The simple fact that he had dealt with non-medical topics sufficed to damage his reputation: "In England, a doctor, if he does not want to lose the confidence of the public, must abstain from engaging in any scientific or literary research which appears to be unrelated to the art of healing."[23] How could a practical physician gain a high reputation, or raise his reputation at all, if he had not always fully dedicated himself to the profession and did not always intend to do so? Young also knew that as a "universal public genius" and an active member of institutions such as the Royal Society, he first had to gain the trust of the public as a physician and should under no circumstances further damage that trust. In the foreword to his 1823 publication on the Egyptian language, he almost defended himself for his (non-anonymous) publication on a non-medical topic: "It may naturally be expected that I should make some apology, for what is generally considered as a violation of professional decorum; for presuming to appear again before the public, without absolute necessity, in any other capacity than that of a practical physician .... It is at least fair [for the public], that direct evidence should be produced or imagined of

22 Thomas Young, *A Reply to the Animadversions of the Edinburgh Reviewers, on some Papers, Published in the Philosophical Transactions*, Longman & Co., London, 1804, p. 37.

23 "En Angleterre, un médecin, s'il ne veut pas perdre la confiance du public, doit s'abstenir de s'occuper de toute recherche scientifique ou littéraire qui semble étrangère à l'art de guérir". Author's translation. François Arago, *Éloge historique du docteur Thomas Young: lu à l'Académie des Sciences, le 26 novembre 1832*, Didot, Paris, 1832, p. 94.

a devotion to medical pursuits, before medical confidence can reasonably be expected."[24]

His reputation as a practical physician did not turn out to be what Young had hoped for throughout his life, as he wrote his friend Hudson Gurney in a letter dated 22 August 1820: "But it has taken me thirty or forty years to become at all known as a man of science and a scholar – it would take me fifty more to be allowed to be a good gentleman and lady's physician."[25] To what extent, then, can Young's status as a practical physician be seen as another factor that caused the failure of his eriometer in the medical context? Could practitioners trust an instrument that was not transparent? One constructed by a physician with little practical experience and a bad reputation as a practical physician?

Despite an inexpensive, easy-to-build, and working instrument, Young's eriometer can be considered a failed instrument in the sense that it failed to gain authoritative status or widespread acceptance among his peers in early nineteenth century England. The reasons for such a failure can be various and complex; they cannot always primarily be attributed to a single factor. A monocausal attempt at explanation would fail too, but in a different way.

---

24 Thomas Young, *An account of some recent discoveries in hieroglyphical literature, and Egyptian antiquities: including the author's original alphabet, as extended by Mr. Champollion, with a translation of five unpublished Greek and Egyptian manuscripts*, John Murray, London, 1823, p. xii.

25 Thomas Young, quoted in Alexander Wood, *Thomas Young: Natural Philosopher 1773–1829*, Cambridge University Press, Cambridge, 1954, p. 77.

CHAPTER 11

Black and White Images in a Grey Area of Failure: the Photo-Theodolite at the Royal Geographical Society in the 1920s

Jane Wess

This paper argues that success or failure of an instrument cannot be judged except in context, and only then with the benefit of hindsight. Instruments do not stand alone, and circumstances allow success or failure to develop over time. A further pertinent consideration is where the success or failure takes place. Starting in the 1830s the Royal Geographical Society in London (RGS) sent instruments out with hundreds of expeditions. Instruments were expected to succeed in allowing travellers to bring back information from an extremely varied set of physical and political situations. Instruments which could have succeeded in less challenging circumstances could fail due to spatial as well as temporal factors. From a wide-ranging study of instruments used on behalf of the RGS, it can be observed that instruments were rarely totally successful or complete failures. However, instances of failure could be minimised in accounts for public consumption.

Following a history and description of the instrument, the paper considers the RGS's acquisition and use in the 1920s of a Wild photo-theodolite, one of the most expensive and eagerly anticipated instruments to be acquired by the institution in its first century. It examines the particular demands of sensitive instruments in the field, which could sometimes stretch the resources of travellers to breaking point. The paper focuses on the expedition of James Alfred Steers, who took the photo-theodolite to the Great Barrier Reef in 1928. The papers of Steers held in the RGS archives relate difficulties not apparent in the published account of the expedition, leading to the assessment of the instrument as a failure in this context.

1 Development of the Photo-Theodolite

By the late 1850s, the establishment of photography as a routine activity encouraged the incorporation of photographic equipment into existing instrument types. One such combination of a traditional instrument and a camera was

 DOI:10.1163/9789004689107_013

the remarkable photoheliograph designed by the British astronomer Warren De La Rue in 1857.[1] In comparison the photo-theodolite had an inauspicious start during the nineteenth century, one reason being theodolites needed to be portable and photographic equipment was cumbersome. Shortly before De La Rue introduced the photoheliograph, Aimé Laussedat was experimenting with photography in military surveying on behalf of the French government.[2] He designed an instrument consisting of a camera with a viewfinder and level.[3] He was joined in the 1850s by others who were independently working with the problems of photographic surveying across Europe, for example, the Italian engineer Ignazio Porro who coined the word "photo-theodolite".[4] In 1859 the Brunner company of Paris produced what has been described as "the first serviceable camera for photographic surveying".[5]

Nevertheless, the photo-theodolite went unremarked and little used. Since Laussedat only published on his photographic methods much later after they had been introduced more widely, his early experiments could be considered failures in that they were not influential.[6] The *Reports* of the United States Commissioners following the Paris Exposition of 1867 were dismissive of the innovations in the theodolites on display.[7] In 1875 Arthur Sherburne Hardy complained that photography was not widely used in surveying in the United States.[8] This is an example of an instrument not being successful long after

1 Science Museum, London, inventory no. 1927–124, Lee T. Macdonald, "'Solar Spot Mania': The Origins and Early Years of Solar Research at Kew Observatory, 1852–1860", *Journal for the History of Astronomy* 46, no. 4 (2015), pp. 469–490.

2 L. Polidori, "On Laussedat's Contribution to the Emergence of Photogrammetry", *International Archives of Photogrammetry, Remote Sensing and Spatial Information Sciences* 43 (2020), pp. 893–899: 893; Thomas Luhmann, Stuart Robson, Stephen Kyle and Jan Boehm, *Close-Range Photogrammetry and 3D Imaging*, 3rd edn., De Gruyter, Berlin, 2020, pp. 19–23.

3 Louis Ragey, "The World of Laussedat and Education in Photogrammetry at the National School of Arts and Crafts, Paris", *Photogrammetric Engineering* 18 (March 1952), pp. 21–26: 23.

4 Olga Chatzifoti, "On the Popularisation of Digital Close-Range Photogrammetry: A Handbook for New Users", MSc thesis, Geoinformatics, National Technical University of Athens, Athens, 2016, p. 22.

5 Gomer T. McNeil, "Metrical Photography", *Photogrammetric Engineering* 23 (1957), pp. 667–677: 669; see also for information about this company, Paolo Brenni, "The Brunners and Paul Gautier", *Bulletin of the Scientific Instrument Society*, no. 49 (1996), pp. 1–8.

6 Aimé Laussedat, *La Métrophotographie*, Gauthier-Villars, Paris, 1899.

7 William P. Blake (ed.), *Reports of the United States Commissioners to the Paris Exposition, 1867*, United States Department of State, Washington, DC, 1870, 6 vols.: v. 2, p. 595.

8 A. S. Hardy, "New Methods in Topological Surveying", *Van Nostrand's Eclectic Engineering Magazine* 12 (1875), pp. 419–426.

it was invented in principle, similar to the situation with the arithmometer which has been described by Stephen Johnston.[9]

It was over ten years later that instruments involving photography became viable in mapping large tracts of land, such as in the Canadian Rockies by J. J. MacArthur, working under Édouard Deville from 1888,[10] reaching 14,000 square miles by 1894.[11] Deville described a number of photographic surveying instruments available in the 1890s.[12] The magazine *Engineering* in 1897 reported that "the idea of using the camera as a surveying instrument has occurred independently to quite a number of engineers", indicating the circumstances were now favourable to this development.[13] One well-known instrument was devised by John Bridges-Lee, who patented a photo-theodolite in 1894, licensing Louis Pascal Casella to make it two years later.[14] By 1899 Casella could report that the instrument had been sold to the United States Geological Survey, the Governments of India, France, Chile, and Mexico, and McGill University, Canada.[15]

An important development was the introduction of the stereoscopic method which took place at the end of the century, but again, sound principles were not sufficient to make an instrument a success.[16] Deville, mentioned above, had designed a stereoscopic photo-theodolite in 1896, but it was inflexible, only being suitable for one viewer.[17] In the first years of the twentieth century stereo-comparators were invented independently by Carl Pulfrich of Zeiss

9 Stephen Johnston, "Making the Arithmometer Count", *Bulletin of the Scientific Instrument Society*, no. 52 (1997), pp. 12–21.

10 Édouard Deville, *Photographic Surveying, including the Elements of Descriptive Geometry and Perspective*, 1st edn., Government Printing Bureau, Ottawa, 1889.

11 Édouard Deville, *Photographic Surveying, including the Elements of Descriptive Geometry and Perspective*, 2nd edn., Government Printing Bureau, Ottawa, 1895, p. v.

12 Deville, *op. cit.* (n. 11), pp. 118–145.

13 "The Bridges-Lee Photo-Theodolite", *Engineering* (10 September 1897), pp. 314–315: illustration p. 312.

14 British Patent no. 14,460 of 1894; London, Archives of the Science Museum, Bridges-Lee Papers, MS 113/5; *Engineering, op. cit.* (n. 13); J. Bridges-Lee, "Photographic Surveying", in John Coles (ed.), *Hints to Travellers Scientific and General*, 8th edn., Royal Geographical Society, London, 1901, 2 vols.: v. 1, pp. 123–128.

15 L. Casella, *Description of the Bridges-Lee New Patent Photo-Theodolite*, 2nd edn., Stationers' Hall, London, 1899, p. 2.

16 Peter Collier, "The Impact on Topographic Mapping of Developments in Land and Air Survey: 1900–1939", *Cartography and Geographic Information Science* 29, no. 3 (2002), pp. 155–174: 158.

17 Édouard Deville, "On the Use of Wheatstone Stereoscope in Photographic Surveying", *Proceedings and Transactions of the Royal Society of Canada*, series 2, 8 (1902), pp. 63–69.

Jena, and Henry Fourcade in South Africa.[18] However, the calculations were extremely complex and time-consuming so the method was only used where others were not possible, such as in Alpine surveying.[19] The situation had not changed fundamentally at the end of the First World War, as most effort was devoted to overcoming problems inherent in the earlier generation of plotters: obtaining co-planar images.[20]

2 The Wild Photo-Theodolites

Nicolàs de Hilster relates how Heinrich Wild, a talented young engineer, became frustrated with the theodolites available in 1902.[21] While working for Carl Zeiss in Jena, Wild listed the desirable features he was hoping to achieve in a theodolite in 1919.[22] He filed for a patent in 1921 which was granted three years later.[23] Also in 1921, he moved to Heerbrugg in Switzerland, in order to set up his own company, forming a collaboration with Dr Robert Hebling from Flums and Jacob Schmidheiny.[24] The following year the Wild company constructed the first commercial photo-theodolite, the Wild P30, and its first photogrammetric plotter, the A1 stereo-autograph, in 1923.[25] The theodolite designed in 1921 and patented in 1924 was the Zeiss Th1, but the Wild T2, a slightly improved version from Wild's own firm, was the one to attract

18 D. J. Lanska, J. M. Lanska, and B. F. Remler, "Carl Pulfrich and the Role of Instruments to Identify and Demonstrate the *Stereo-Effekt*", *Neurosciences and History* 3, no. 1 (2015), pp. 8–18; Henry G. Fourcade, "A Stereoscopic Method of Photographic Surveying", *Nature* 66 (1902), pp. 139–141.

19 George D. Whitmore and Morris M. Thomson, "Photogrammetry Comes into Its Own", *Photogrammetric Engineering* 19 (1953), pp. 21–30: 24–25; Collier, *op. cit.* (n. 16), p. 158.

20 Collier, *op. cit.* (n. 16), p. 166.

21 Nicolàs de Hilster, "Geodetic Instruments: Wild Heerbrugg", https://www.dehilster.info/geodetic_instruments/wild_heerbrugg.php (accessed 20 March 2023).

22 Heinrich Wild, "Der neue Theodolit", *Schweizerischen Zeitschrift für Vermessungswesen und Kulturtechnik* – 12 May 1925; from J. Dedual of the Virtual Archive of Wild-Heerbrugg, 2015, https://wild-heerbrugg.ch/heinrich-wild/wild-der-neue-theodolit (accessed 22 March 2023).

23 United States Patent 1,508,585, filed 13 May 1921, granted 16 September 1924; see Google Patents, https://patents.google.com/patent/US1508585A/en (accessed 20 March 2023).

24 Bernhard Braunecker, "Werdegang der industriellen Optikentwicklung in Heerbrugg", *Swiss Physical Society*, September 2011, see https://www.sps.ch/en/artikel/physik-anekdoten/werdeg (accessed 8 May 2023); Otto Stockmaier, *Leica Heerbrugg im Wandel der Zeit*, Leica AG, Heerbrugg, Switzerland, 1996; Nicolàs de Hilster, "Geodetic Instruments: 1924 Carl Zeiss Th1 Theodolite", https://www.dehilster.info.geodetic_instruments/1924_carl_zeiss_jena_th1_theodolite.php (accessed 20 March 2023).

25 Collier, *op. cit.* (n. 16), p. 166.

attention.[26] "Considerable excitement in scientific and survey circles" was caused by the introduction of the T2 theodolite by Heinrich Wild in 1924, which was seen as ushering in a new generation of instruments.[27]

From Wild's original list of desiderata, from the patent, and also from E. R. L. Peake's paper of 1929, which gives a meticulous history of Wild's development of his theodolite, the attributes required by Wild are known.[28] The Zeiss Th1 was intended to meet the following requirements: smallest possible dimensions and weight; convenient manipulation; greater robustness in respect of transport, rain, and dust; combination of images from two opposite parts of the circle on one fine dividing line for observation in one eyepiece; images of opposite points of the circle to be reproduced through the hollow axis, so that the circles might be completely enclosed; the use of a highly sensitive micrometer for adjustment of coincidence to allow direct reading of the arithmetical mean; and the use of glass circles to ensure a symmetrical graduation.[29] The Wild T2 also had an anallactic telescope of constant length with internal focusing (which obviated the need to add the focal length to the observed distance from a stadia), a device for centring the bubble by making apparent coincidence of its two ends as viewed in a prism, and a cylindrical steel main axis.[30]

The heralding of these theodolites as new instruments by writers such as D. A. Hutchison and Arthur Hinks was justified given their considerable improvements in accuracy, manipulation and protection from the weather.[31] Peake recounted that Canadian surveyors found the Wild T2 with 3.75-inch circles was equal in accuracy to a 12-inch traditional design.[32] From the interest in the Wild instruments it can be argued that they stimulated an increase in the use of stereoscopic photographic surveying. However, their success depended crucially on a backup system which in some localities was not available.

26 Nicolàs de Hilster, "1929 Wild Heerbrugg T2", https://www.dehilster.info/gedetic_instruments/1929_wild_heerbrugg_t2_theodolite.php (accessed 25 March 2023).

27 Collier, *op. cit.* (n. 16), p. 167.

28 Wild, *op. cit.* (n. 22); United States Patent, *op. cit.* (n. 23); E. R. L. Peake, "The Tavistock Theodolite: A Paper Read at the Afternoon Meeting of the Society on 11 February 1929", *Geographical Journal* 73, no. 6 (June, 1929), pp. 513–519.

29 Peake, *op. cit.* (n. 28), p. 513.

30 O. von Gruber, "Neue Theodolitkonstruktionen", in leaflet: *GEO 57, Carl Zeiss Jena* (1927), pp. 1–8: 8, as cited by de Hilster, *op. cit.* (n. 26).

31 D. A. Hutchison and Arthur Hinks, "The Photo-Theodolite of Mr H. Wild", *Geographical Journal* 67, no. 3 (March, 1926), pp. 233–240.

32 Peake, *op. cit.* (n. 28), p. 513; Anita McConnell, *Instrument Makers to the World: A History of Cooke, Troughton and Simms*, William Sessions, York, 1992, pp. 79–80.

3 The Photo-Theodolite at the RGS: Purchase and Celebration

The first indication of interest on behalf of the RGS was in connection with a reconnaissance expedition to investigate whether Mount Everest could be climbed in 1921.[33] The Research Committee decided to ask the Survey of India for a photo-theodolite, and have the results worked up in Paris on the Orel apparatus owned by the Société Française Stéréotopographie.[34] Eduard von Orel was an Austrian Lieutenant who designed a method of producing contour maps from stereophotographs before the First World War.[35] Known as the "stereoautograph", the Orel plotter was significant in that it allowed operators to trace elevation contours directly, avoiding some of the computation required by other plotters.[36] Two papers about the instrument (*Fig. 11.1*) were published by the RGS in 1922, followed by a discussion.[37] Hinks revealed a reason why an instrument might not be a success: "The policy of the makers is to grant a license for the sole use of the apparatus in any one country ... this method of exploitation has disagreeable features"; and "it seems to preclude any use except commercially for profit". Hence nothing became of the initiative on behalf of the RGS for the Everest expedition.

The RGS Research Committee again expressed interest in the "photo-mechanical process" towards the end of 1924, and in 1925 discussed the photo-theodolite by Wild in some detail.[38] It was reported that "the new apparatus by Wild" had been tried during two seasons in the Alps and was "probably the best and certainly the least expensive of instruments of its kind". Together with accessories it cost £305.[39] However, the essential stereoautograph, sometimes referred to as a "stereocartograph", was in use in Switzerland. This apparatus

33 Expedition Committee Minutes, 10 January 1921: London, Royal Geographical Society-Institute of British Geographers (hereafter RGS-IBG), Committee Minutes, 1917–1926, p. 102; Hutchison and Hinks, *op. cit.* (n. 31), p. 240.

34 Research Committee Minutes, 14 March 1921: London, RGS-IBG, Committee Minutes, 1917–1926, p. 112.

35 Arthur W. Judge, *Stereoscopic Photography: Its Application to Science, Industry and Education*, 2nd edn., Chapman and Hall, London, 1935, pp. 294–296.

36 Collier, *op. cit.* (n. 16), p. 158.

37 Arthur R. Hinks, "Stereographic Survey: The Stereoautograph", *Geographical Journal* 59, no. 4 (April 1922), pp. 273–284: 279; N. M. MacLeod, "Stereographic Survey: The Autocartograph", *ibid.*, pp. 284–289; E. M. Jack, G. T. McCaw, and Major Henrici, "Stereographic Survey: The Autocartograph: Discussion", *ibid.*, pp. 289–292.

38 Research Committee Minutes, 10 November 1924: London, RGS-IBG, Committee Minutes, 1917–1926, p. 230; Expedition Committee Minutes, 9 November 1925: London, RGS-IBG, Committee Minutes, 1917–1926, p. 208.

39 "Catalogue of Instruments", 1879–1933: London, RGS-IBG, Additional Papers, AP 52, pp. 2A, 49B.

THE STEREOAUTOGRAPH OF VON OREL AND ZEISS
By courtesy of Colonel Talon.

FIGURE 11.1 The Stereoautograph of Von Orel and Zeiss, 1922, from Arthur R. Hinks, "Stereographic Survey: The Stereoautograph", *Geographical Journal* 59, no. 4 (April 1922), between pp. 276 and 277

cost upwards of £2,400 and was well beyond the Society's budget. The decision was made to purchase the photo-theodolite with the expectation that the RGS would be able to use the stereoautograph when it became free. This was one of the tentative approaches on behalf of the RGS to forging international links after the First World War.

The Research Committee discussed the photo-theodolite again on 14 December 1925, proposing to show it at the next Council meeting.[40] However, by 18 January 1926 the photo-theodolite had arrived but was not on exhibition.[41] A technical paper describing it in glowing terms by D. A. Hutchison and A. R. Hinks, the latter of whom was secretary for thirty years from 1915, was produced rapidly, appearing in the *Journal* for March 1926.[42] The authors noted that the Wild theodolite and photo-theodolite were not well-known in England, and therefore offered a detailed description and photographs

40 Research Committee Minutes, 14 December 1925: London, RGS-IBG, Committee Minutes, 1917–1926, p. 287.

41 Research Committee Minutes, 18 January 1926: London, RGS-IBG, Committee Minutes, 1917–1926, p. 295.

42 Hutchison and Hinks, *op. cit.* (n. 31).

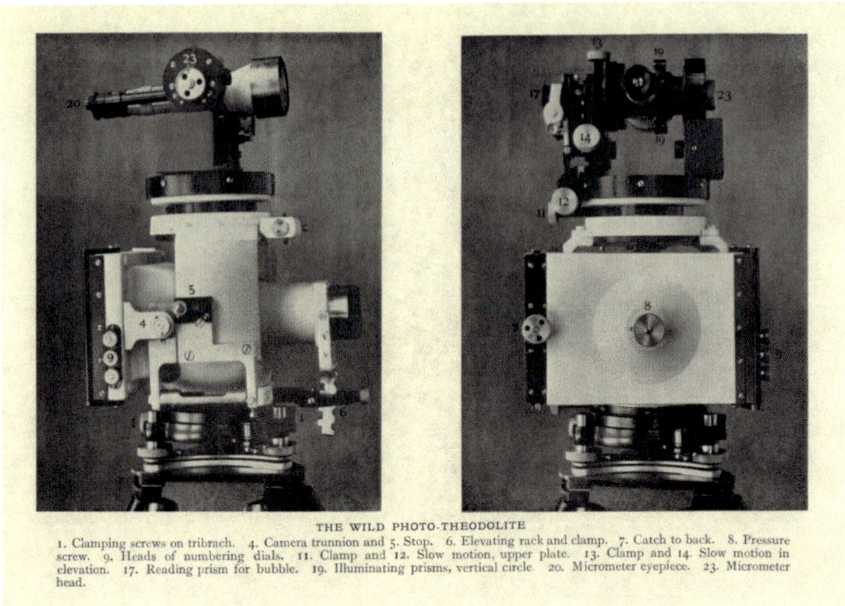

FIGURE 11.2 Photograph of the Wild photo-theodolite acquired by the Royal Geographical Society in 1926. From D. A. Hutchison and Arthur Hinks, "The Photo-Theodolite of Mr H. Wild", *Geographical Journal* 67, no. 3, March 1926, between pp. 236 and 237

(*Fig. 11.2*). This lack of a British equivalent was keenly felt, leading to a trial of the Wild theodolite at Tavistock in 1926 on behalf of the Admiralty, War Office and Ordnance Survey, where British instrument makers were encouraged to construct a lightweight instrument to compete with that of Wild.[43]

Hutchison and Hinks moved from a description of the Wild theodolite to the photo-theodolite (*Fig. 11.2*): "It is an ordinary theodolite with a camera inserted between the horizontal plates. This was made possible by using circular ball races instead of the normal bearing."[44] Two cameras were provided, with focal lengths of "about 16cm and 25cm, with lenses of a special type computed by Mr Wild, and fitted with yellow screens of factor ×4". The plates were coated on glass and held against the camera edges, and against "pierced plates acting as fiducial marks, and with small engraved glass drums which print serial numbers on the plates". While the paper was generally approving, the instrument was said to be a "master-piece of the instrument maker's art", although "one or

43 Peake, *op. cit.* (n. 28), p. 513.

44 Hutchison and Hinks, *op. cit.* (n. 31), p. 238.

two defects were noted such as light reflection from the sky or from the filament lamp used at night."[45]

One problem with apparatus so reliant on visibility became apparent very soon. Bad weather prevented a trial: "On Wednesday it was taken to the Hog's Back near Guildford but visibility was so poor that all hopes of giving the instrument a satisfactory test were abandoned."[46] This first setback was one not confined to the photo-theodolite, but affected its use more than that of other surveying instruments.

4 Initial Demand

Since the photo-theodolite had such a positive reputation there followed a flurry of requests for its use. During the First World War the government had commandeered the resources of the RGS, and this relationship was again apparent regarding the allocation of the photo-theodolite. The government requisitioned the instrument for the Shaksgam Valley expedition, then known as the North West India "frontier". This expedition was the first recorded use of the photo-theodolite. The results of the 1926 expedition were recounted by the explorer and surveyor Kenneth Mason.[47] Mason admitted he had some problems with the instrument in the field. At times, he had to resort to plane-tabling, and complained of the photographic plates getting spoiled by light as they could not be developed immediately; they were due to be worked up in Flums, Switzerland, by the stereoautograph the following spring.[48] However the publication gave no indication of the difficulties that were later revealed by the Great Barrier Reef expedition.

On 12 April 1926 the RGS Council decided that the photo-theodolite should go to the Singalila Ridge after it returned from Shaksam. However, the Surveyor General of India wanted the instrument for use in Iraq for a survey of the Mosul border.[49] The RGS responded by saying that it would lend it if it was properly insured and if the Society would be allowed to explore the Himalayas

45 Hutchison and Hinks, *op. cit.* (n. 31), p. 236.

46 Hutchison and Hinks, *op. cit.* (n. 31), p. 238.

47 Kenneth Mason, *Exploration of the Shaksam Valley and Aghil Ranges 1926*, Surveyor General of India, Dehra Dun, India, 1928.

48 Expedition Committee Minutes, 9 November 1926: London, RGS-IBG, Committee Minutes, 1926–1929, p. 33.

49 Expedition Committee Minutes, 25 October 1926: London, RGS-IBG, Committee Minutes, 1926–1929, p. 24; *International Boundary Study No. 27 Iraq-Turkey Boundary*, Office of the Geographer in the Bureau of Intelligence and Research, Washington, DC, 1964, p. 5.

in future.[50] It can be seen that the instrument had agency beyond providing measurements in that it could act as a bargaining chip to open up politically difficult territory. In this case all these negotiations were brought to a halt as the Council minutes for 22 November state baldly that the War Office would be using the photo-theodolite.[51] The RGS merely asked that any results should be disseminated. The Mosul boundary project took up most of 1927, but other contenders followed. In March 1928 Richard van R. Woolley asked for the instrument to go to Drakensberg in Natal, which was granted, and in February 1929 Frank Debenham asked for it to go to an unspecified location, which was also granted.[52]

The RGS "Catalogue of Instruments" is slightly confusing as it lists two photo-theodolites as having been purchased from Wild in 1926. In fact, this was one instrument which had been provided with two cameras as described above. The cameras appear to have been separated before Alfred Steers went to Australia. A note by Woolley remarks that he took only one camera.[53] Possibly because of the demand for the photo-theodolite, the RGS constructed a second photo-theodolite from an older British theodolite and a Wild camera in a manner typical of the then rather tight-fisted institution.

5 Expedition to the Great Barrier Reef

At the time that "photo-theodolite 1" was in Drakensberg, "photo-theodolite 2" was assigned to the Great Barrier Reef, for which expedition Steers applied in 1928.[54] Steers was now better-known as an academic geographer, particularly interested in coastal features, who built the department at Cambridge into a leading centre for physical geography. There are conflicting views of this expedition. Steers's obituary describes it as "perhaps the most outstandingly

50 Expedition Committee Minutes, 26 October 1926: London, RGS-IBG, Committee Minutes, 1926–1929, p. 32.

51 Council Minutes, 22 November 1926: London, RGS-IBG, Council Minutes, v. 16, p. 74.

52 Expedition Committee Minutes, 5 March 1928: London, RGS-IBG, Committee Minutes, 1926–1929, p. 108; Council Minutes, 18 February 1929: London, RGS-IBG, Council Minutes, v. 17, p. 246. Richard van der Reit Woolley subsequently became the eleventh Astronomer Royal; Frank Debenham was the first director of the Scott Polar Institute in Cambridge.

53 "Catalogue of Instruments", 1879–1933: London, RGS-IBG, Additional Papers, AP 52, pp. 2A, 49B.

54 James Alfred Steers, "Diary Great Barrier Reef expedition": London, RGS-IBG, James Alfred Steers Collection, JAS/1.

successful and influential collaborative field project of its time".[55] On the other hand, more recently it has been described as a "fiasco".[56] Certainly it is difficult to see results in comparison to the resources invested, and the photo-theodolite did not play a part in any successes there may have been.

The expedition came early in Steers's career. His resulting paper was read at the evening meeting of 4 February 1929 and published in two parts in the *Geographical Journal* later that year.[57] The expedition was originally set up to investigate certain biological problems, including the reef's formation. It was planned by the Brisbane Barrier Reef Committee led by Sir Matthew Nathan, who had also formed an English committee to act in conjunction with it.[58] There were hopes that the reef could potentially become lucrative by benefitting the pearl, turtle and sponge industries.[59] Steers reported that the expedition had a Geographical Section under the auspices of the RGS, which had provided several surveying instruments.[60] The RGS had not wanted to be restricted to one area. Steers and his assistant, M. A. Spender, had made an extensive reconnaissance of the Queensland Coast between Whitsunday Islands and Flinders Island, and had visited the extreme southern end of the reef.[61] The publication consisted of description, speculation, and argument, but did not mention his methods.

The manuscript diary that Steers kept during the voyage is far more revealing. The remainder of this section is taken from that document.[62] They set off on 14 June 1928 for an extensive tour of the coast of Australia, returning to England on 10 January 1929. Steers was not naturally inclined to instrumental measurement. The expedition had been given "a large thermometer" on which Steers recorded 105 degrees Fahrenheit on 28 June in the Red Sea off the coast of Sudan on the journey out. Travellers possessing an instrumental approach took temperature and various other measurements at regular intervals. Those more concerned with evoking their surroundings focused on recording extremes. Steers fell into the latter category. He also used compass

55 "Obituary: James Alfred Steers, CBE, 1899–1987", *Geographical Journal* 153, no. 3 (1987), pp. 436–438.

56 "Obituary" *op. cit.* (n. 55); James Bowen and Margarita Bowen, *The Great Barrier Reef: History, Science, Heritage*, Cambridge University Press, Cambridge, 2002, p. 274.

57 J. A. Steers, "The Queensland Coast and the Great Barrier Reefs", *Geographical Journal* 74, no. 3 (September 1929), pp. 232–257; J. A. Steers, "The Queensland Coast and the Great Barrier Reefs (continued)", *Geographical Journal* 74, no. 4 (October 1929), pp. 341–367.

58 Steers, *op. cit.* (n. 57), p. 232.

59 *Sydney Morning Herald*, 27 June 1928, p. 14.

60 Steers, *op. cit.* (n. 57), p. 232.

61 Steers, *op. cit.* (n. 57), p. 232.

62 Steers, "Diary", *op. cit.* (n. 54).

directions rather than degrees which exposes a lack of affinity with numerical measurement. At one point he gave a rough estimate of height above sea level where some travellers would have taken measurements with a hypsometer.[63]

6 Steers's Diary

The diary records a great deal of photographic activity, and much accompanying concern. On 30 June 1928 Steers discovered a defect in the camera which made him "much worried". The following day it seemed to work with parallel plates. They reached Socotra, an island off Yemen, in a heavy sea where he sent off a wireless message regarding the camera on 2 July. The two stereo plates taken of Gibraltar were "bad failures" but on 7 July Steers took several more and "hoped the stereo camera would work". On 12 July he worked at his photo survey for "a little in the afternoon". On 18 July they arrived at Fremantle, Western Australia, and on 27 July they docked at Sydney, New South Wales, where he took the field glasses to be repaired. The following day he took the camera to Kodak "who are dealing with it thoroughly and I fear expensively!" However, when he called for the camera on 1 August the reflex lever broke. Fortunately, the next day he retrieved it and "all seems well and cost very little". The slides from Ceylon were "not too bad".

Having recently run into a mud bank through "careless navigation" on 16 August, Steers took "a good many photographs" (*Fig. 11.3*). On 20 August, Kodak exposed films but they were "fair only"; he had overexposed in many cases. On 24 August he took some stereopairs and the next day took them to be developed: "We are now quite worried about the slide-plate carrier of the photo-theodolite which has not yet emerged from the cases and may not have been sent". On 27 August he recounted that "The photos taken yesterday were not a success, somehow or other they were blurred and two of them mis-fired altogether. Two tiny holes found in the shutter. I do hope the camera is all right". He spent most of the day in Kodak's premises. On 30 and 31 August they made "a rather careful reconnaissance for a photo survey. I am worried about results – but best not to think about that at this stage".

At the start of September, Steers began to rebel against the instrumental regime which he felt was getting in the way of him exploring the reefs. On 1 September he wrote: "many small difficulties arise chiefly [from] lack of contact points which can be picked out by the camera". Steers now revealed his

63 Steers, "Diary" *op. cit.* (n. 54), 8 September 1928. As Steers's "Diary" is unpaginated, references can be found under the individual date cited.

FIGURE 11.3 Hope Islands, mangrove and shingle cay, photograph by M. A. Spender, 1928. The Hope Islands are at the north end of the Great Barrier Reef
ROYAL GEOGRAPHICAL SOCIETY, 057170. © ROYAL GEOGRAPHICAL SOCIETY (WITH THE INSTITUTE OF BRITISH GEOGRAPHERS)

approach to the work, which, like some other geographers of the early twentieth century, was at odds with the RGS's rigid attitude to mathematical survey: "I am most anxious to see as much of the coast as I can. The more I think about the reef the more I am convinced one needs to think in general or abstract lines rather than by adopting survey methods to certain areas". However, Steers persevered. The following day, they were "busy with the photo-theodolite but we were very heavily laden and it was all that we could do to get the stuff from place to place". Steers had now decided that he must do the survey work even though "it means cutting down on activities very much and also preventing our seeing the coastal features in general". On 3 September cattle had destroyed a beacon. However, "all the survey plates were good and stereos not too bad". He had made an error in focusing in a couple of images. During 8 and 9 September they took photographs with the photo-theodolite and on 11 September they went to Cardwell, Queensland, to refit.

Sailing near Goold Island on 12 September, Steers was beset by physical problems. When discussing plans for the next few days they decided "a photo survey was out of the question as the going was very rough .... We made a tacheometric-cum-plane table map of the Dunk Spit, Richard Island photographed with no small difficulty" on 18 September. The following few days they continued surveying the area using a tacheometer and plane table. On 21 September they surveyed Beaver Cay in the morning, again with a tacheometer, levelling staff and plane table: "This proved a quick and accurate method". It appears that in order to get the surveying done in the most efficient and secure manner, Steers had fallen back on earlier and simpler methods.

They were now distinguishing between "official" and "non-official" photographs. Presumably the official ones were those demanded by the RGS, and the others were ones that Steers considered interesting. On 23 September Steers wrote: "Rather steep slopes made carrying apparatus rather risky. Spender got stuck and I must not risk the apparatus in this way, for photographs that are not part of the photo-survey proper". They began to have more success. On 21 September they developed "official" photographs, which were excellent. On 5 October they developed plates which were good. On 14 October they took "official" photographs of hills that Steers hoped would be as good, as "the stereo effect was particular". During the second half of October they took many photographs as well as continuing with tacheometric surveys. Occasionally stereophotography was particularly mentioned, as on 9 November.

On 30 November 1928 Steers went to Kodak in Sydney and "saw nearly all my slides … I felt satisfied with these … Kodak's people seemed to suggest we did well" (*Fig. 11.4*). Another batch of plates were taken in on 3 December. Large numbers were collected on 6 and 7 December, and altogether Steers was "fairly pleased". On 8 December the last load was collected and packed, and they left Sydney for home.

7 Conclusions

Given the amount of time and effort spent on the photographic survey with the photo-theodolite, followed by the development of the plates, it is surprising that no mention of this aspect of the expedition was mentioned by Steers in what became established knowledge.[64] In the second of his papers there were no stereoscopic pairs of photographs, nor any mention of the photo-theodolite, which focused instead on his ideas concerning the formation of the reefs. The photo-theodolite was not mentioned in a single newspaper article written concerning the expedition in Australia.[65] Steers's personal interest in

64 Steers, *op. cit.* (n. 57).

65 See National Library of Australia website: https://trove.nla.gov.au/newspaper/ (accessed 25 March 2023).

FIGURE 11.4 Kodak's shop front, George Street, Sydney, 1932–1934
MUSEUMS VICTORIA, GLASS NEGATIVE, HTTPS://COLLECTIONS.MUSEUMSVICTORIA.COM.AU/ITEMS/2067172. COURTESY OF KODAK (AUSTRALASIA) PTY LTD

physical geography, and his evident lack of enthusiasm for surveying was partly responsible, but so too was the difficulty of using the apparatus and the issues surrounding its fragility and maintenance. There seems no doubt that photo-theodolites became increasingly popular and successful, but not in the period under discussion. In 1932 a publication describing the instrument appeared, suggesting it was not familiar.[66] Only in the years following the Second World War did the instrument come into its own. The following quote from 1953 makes the point succinctly: "Only in the last decade or so has photogrammetry had a really profound influence on the economies of surveying and mapping. Within the last ten years photogrammetry instead of field work with the plane

66 Clinton Gresham Lewis, "The Wild photo-theodolite in triangulation and in stereographic surveys from photographs taken from the ground", in Desmond Roe Crone, *Instructions for the Use of the Wild Universal Theodolite*, Survey of India Departmental Paper No. 14, Dehra Dun, 1932, pp. 9–26.

table has become the basic procedure for the making of topographic maps and for many other activities in the surveying field".[67] Wild photo-theodolites, together with those made by Zeiss, were the market leaders.[68]

However, the limited information accessible for expeditions in the late 1920s does not support the view that photo-theodolites were successful then, especially not in places where logistical backup was not immediately available. Published accounts of expeditions at that time tended to minimise any difficulties, but the diary of Alfred Steers at the Great Barrier Reef in 1928 makes the frustrations apparent. Even allowing for the fact that this expedition did not benefit from the complete instrument provided by Wild, in that only one of the two cameras provided was taken, the camera was liable to faults and appears to have been too fragile and cumbersome for the requirements of a travelling expedition. Numerous problems with both the quality of the plates and the suitability of the terrain meant that Steers ultimately fell back on simpler methods which could be undertaken faster and more efficiently. Except in Sydney, days away from much of the reef, there was no infrastructure to support the use of this relatively sophisticated piece of equipment.

While many of Steers's difficulties were caused by the weight of his equipment, an additional problem was the cost of this very expensive and novel camera, valued at £98 10s.[69] The awareness of the cost as well as the fragility only served to worry Steers, further inhibiting him from using the instrument to advantage. When historians allocate the labels "success" or "failure" to instruments, it must be accompanied by a recognition of all the factors in play. An instrument which works well in theory may not be suitable at different times or in different terrains. As the context varies, so must the assessment of any instrument.

Acknowledgements

I would like to thank Nicolàs de Hilster for providing several original sources for the Wild instruments in his web pages, and Julian Holland for information on the Great Barrier Reef expedition in Australian newspapers.

---

67 Whitmore and Thomson, *op. cit.* (n. 19), p. 28.

68 Collier, *op. cit.* (n. 16), pp. 168–169; Heinz Gruner, "Photogrammetry: 1776–1976", *Photogrammetric Engineering and Remote Sensing* 43, no. 5 (May 1977), pp. 569–574: 571–572.

69 "Catalogue of Instruments", 1879–1933: London, RGS-IBG, Additional Papers, AP 52, pp. 2A, 49B.

CHAPTER 12

A Failure, an Icon or Beyond: the Cloud Chamber of the University of Padua

Sofia Talas and Giulio Peruzzi

Within his late nineteenth-century studies on the formation of fog, the Cambridge scientist Charles T. R. Wilson brought pure vapour to temperatures below the temperature of condensation without condensing.[1] In this unstable state, obtained by a quick expansion, any impurity represented a condensation nucleus and, in 1911, Wilson developed an effective instrument using this phenomenon to visualise particles: as a charged particle passed through a gas and ionised it, its track was seen as a thick row of drops.[2] The new devices, called "cloud chambers", started to be widely used to observe and photograph particles' tracks.

The history of cloud chambers, their technical evolution and the possible interpretations of their developments and uses have been widely studied from the mid-twentieth century.[3] In this paper, we will focus on the cloud chamber housed at the Giovanni Poleni Museum of the History of Physics, in Padua. This device was built in the late 1930s for Bruno Rossi, who was then professor of Experimental Physics at the University of Padua. We will examine this instrument's biography and discuss the possible ways to interpret this cloud chamber today, in particular in a museum context.

1 See for instance Charles T. R. Wilson, "Condensation of water vapour in the presence of dust free air and other gases", *Philosophical Transactions of the Royal Society* 189 (1897), pp. 265–307.

2 Charles T. R. Wilson, "On a Method of Making Visible the Paths of Ionising Particles through a Gas", *Proceedings of the Royal Society* 85 (1911), pp 285–88; "On an Expansion Apparatus for Making Visible the Tracks of Ionising Particles in Gases and Some Results Obtained by Its Use", *Proceedings of the Royal Society* 87 (1912), pp. 277–292.

3 For an overview on the topic, see Eugenio Bertozzi, "Establishing and Consolidating a Research Field: The Bibliography of the Wilson Cloud Chamber in the History of Particle Physics", *Historical Studies in the Natural Sciences* 49 (2019), pp. 117–150. For a more technical analysis, see Niraj Nath Dasgupta and S. K. Ghosh "A Report on the Wilson Cloud Chamber and Its Applications in Physics", *Reviews of Modern Physics* 18 (1946), pp. 225–290.

 | DOI:10.1163/9789004689107_014

1 The Padua Cloud Chamber: Its First Life, before the Second World War

Bruno Rossi (1905–1993) is today regarded as one of the great physicists of the twentieth century.[4] He was a pioneer of cosmic-ray researches, as he got interested in the topic in 1929, when a lively debate about the nature of cosmic rays opened up: before entering the atmosphere, are cosmic rays charged particles or very high energy electromagnetic waves? Rossi, who worked at the time at the University of Florence, became one of the main protagonists of the debate: he designed new instruments, discovered new properties of the cosmic radiation and proposed, in 1930, a way to solve the controversy, which led to the discovery that primary cosmic rays mainly consisted of positive charged particles.[5]

In 1932, Bruno Rossi was appointed to the chair of Experimental Physics at the University of Padua. There he introduced his experimental research on cosmic rays and, from 1934, he planned and supervised the construction of a new building for the university's Institute of Physics.[6] This new cutting-edge Institute was inaugurated in the spring of 1937 (*Fig. 12.1*).[7] It had workshops, large laboratories for students' training, and fifty research laboratory spaces. The Institute's tower was to be used for cosmic-ray research and two high-voltage generators were built. One of these was a 1-million-volt Cockcroft-Walton generator to accelerate ions and study nuclear disintegrations (*Fig. 12.2*). The generator was fully working in 1938, while the acceleration tube was still under construction. Only one other accelerator – also a 1-million-volt Cockcroft-Walton – was being built in those years in Italy, at the Istituto Superiore di Sanità in Rome; it started working at the end of 1939.[8] To deflect particles and study their properties, Rossi also had a 7-ton electromagnet built, which was ready by 1937.

---

4 See for instance A. Pascolini (ed.), *The Scientific Legacy of Bruno Rossi*, Imprimenda, Università degli Studi di Padova, Padua, 2006; Bruno Rossi, *L'enigma dei raggi cosmici*, Montedison, Progetto Cultura, Milan, 1986 (new edn.: Padova University Press, Padua, 2012).

5 About Bruno Rossi's works during his Italian years, see also Giulio Peruzzi and Sofia Talas, "The Italian Contributions to Cosmic-Ray Physics from Bruno Rossi to the G-Stack: A New Window into the Inexhaustible Wealth of Nature", *La rivista del Nuovo Cimento* 30 (2007), pp. 197–257.

6 Up to then, physics had been taught in the Palazzo Bo, the fifteenth-century building of the University.

7 Bruno Rossi, "Il nuovo Istituto di Fisica della R. Università di Padova", *La Ricerca Scientifica* (1938), pp. 220–227; Angelo Drigo, "L'Istituto di Fisica della R. Università di Padova", *L'Energia Elettrica* 16 (1939), pp. 3–20.

8 Giovanni Battimelli, "Le origini del laboratorio di fisica", *Rendiconti della Accademia nazionale delle scienze detta dei XL* (1999), pp. 149–160. The paper is part of the proceedings of

FIGURE 12.1 The newly built Institute of Physics in Padua, 1937
PHOTO COURTESY OF THE GIOVANNI POLENI MUSEUM, UNIVERSITY OF PADUA

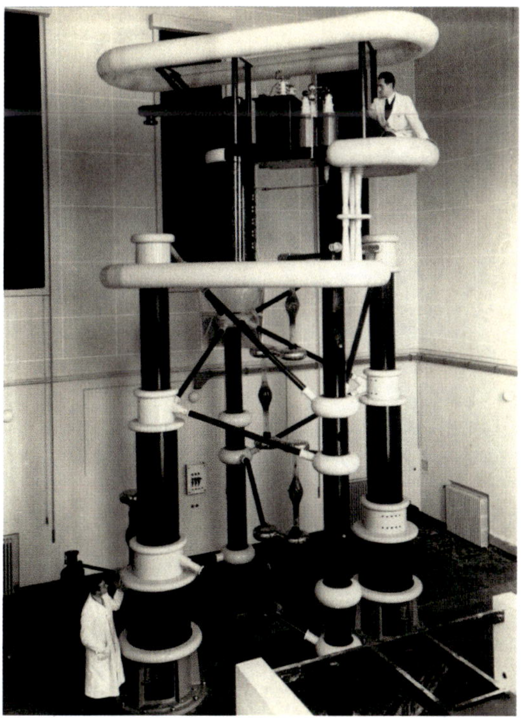

FIGURE 12.2
The Cockcroft-Walton generator set up at Padua Institute of Physics, 1938
PHOTO COURTESY OF THE UNIVERSITY OF PADUA'S ARCHIVES

Designed by the engineer Giovanni Someda, it was regarded as one of the most efficient instruments of this kind in Europe.[9] Up to that time, Rossi had been performing his experiments on cosmic rays only with Geiger-Müller counters but, in the late 1930s, he had a cloud chamber made as well.

Cloud chambers had become fully effective in cosmic-ray researches from the early 1930s, thanks to two main improvements. Firstly, in 1932, Patrick M. S. Blackett and Giuseppe Occhialini, at the Cavendish Laboratory in Cambridge, applied to cloud chambers the coincidence circuit designed by Bruno Rossi in 1930. Made of triodes and Geiger-Müller counters, the coincidence circuit was an AND logic gate (which uses two or more inputs to generate one output): by placing one counter above the chamber and another one below, the expansion mechanism and the photographic camera were triggered only when charged particles passed through both Geiger-Müller tubes. Cosmic-ray particles could thus "take their own cloud photographs", as Blackett and Occhialini wrote in 1933.[10] Furthermore, C. T. R. Wilson proposed a new model of cloud chamber in 1933, which was simpler to construct and could be used in any position, horizontal or vertical.[11] Up to that moment, in cloud chambers, a volume change was produced by the quick motion of a piston, which formed the "floor" of the chamber and was brought to rest by hitting a base plate. Wilson's new cloud chamber had a fixed porous wall of wire gauze, which allowed a portion of the gas to escape when an elastic membrane moved. This membrane was attached to a second chamber separated from the cloud chamber by the gauze partition. This new apparatus was quickly adopted, in particular in cosmic-ray research.[12]

Rossi's cloud chamber, which is now kept at the Giovanni Poleni Museum in Padua, is a refined apparatus intended for research, certainly made by the workshop of the Institute of Physics (*Fig. 12.3*).[13] We did not find any descrip-

the *Convegno in onore di Domenico Marotta nel 25. anniversario della morte*, published in the *Rendiconti*, pp. 79–247.

9 Giovanni Someda, "L'elettromagnete dell'Istituto di Fisica della R. Università di Padova", *L'elettrotecnica* 24 (1937), pp. 470–474.

10 Patrick M. S. Blackett and Giuseppe P. S. Occhialini, "Some Photographs of the Tracks of Penetrating Radiation", *Proceedings of the Royal Society* 139 (1933), pp. 699–727: 699.

11 Charles T. R. Wilson, "On a New Type of Expansion Apparatus", *Proceedings of the Royal Society of London* 142 (1933), pp. 88–91.

12 Blackett for instance started using this new cloud chamber from 1934; see: Patrick M. S. Blackett, "The Measurement of the Energy of Cosmic Rays. I. The Electro-Magnet and Cloud Chamber", *Proceedings of the Royal Society of London* 154 (1936), pp. 564–573.

13 The workshop had been making and repairing instruments for Padua University's Cabinet of Physics from the early nineteenth century, under the supervision of the various professors of Experimental Physics. It made all the cutting-edge instruments needed for physics research that were not easily available on the market.

FIGURE 12.3 The cloud chamber kept at the Giovanni Poleni Museum in Padua
PHOTO COURTESY OF THE GIOVANNI POLENI MUSEUM, UNIVERSITY OF PADUA

tion of the cloud chamber in Rossi's papers, and only a vague reference to the possible use of a cloud chamber is provided in 1938 by one of Rossi's collaborators, Angelo Drigo.[14] However, we know for sure that Rossi had built the cloud chamber kept at the Poleni Museum: we will describe the cloud chamber and further discuss its attribution in the second section of this paper.

At the time it was built, the Padua cloud chamber was one of the first cloud chambers in an Italian laboratory, after the two examples acquired in Rome by Enrico Fermi and his collaborators in 1933.[15] In Milan, the construction of

14 Drigo underlined how the 7-ton electromagnet perfectly fitted the objectives of Padua physicists, who were interested in studying the curvature of particle tracks "for instance through Wilson cloud chamber"; see: Drigo, *op. cit.* (n. 7), p. 19. This is of course quite vague, as Drigo does not clearly refer to an existing cloud chamber.

15 The cloud chambers were designed and made under the supervision of one of the members of Fermi's group, Franco Rasetti, who had learned about these techniques in

a cloud chamber started in 1939 and the device was ready in July of that same year, but the local physicists finished dealing with the chamber's illumination and events-registration issues only in 1941.[16]

The fact that Bruno Rossi never described nor mentioned his cloud chamber is probably due to the fact that he barely had time to use it: in September 1938, Rossi was forced to leave the University of Padua because of the fascist racial laws. After short periods in Copenhagen and Manchester, respectively invited by Niels Bohr and Patrick Blackett, Rossi migrated to the United States, working at first in Chicago, invited by Arthur Compton, and later at Cornell University, invited by Hans Bethe. He then moved to Los Alamos, to collaborate on the Manhattan project, and finally, from 1946 onwards, he worked at the Massachusetts Institute of Technology.

2 The Second Life of the Padua Cloud Chamber, after the Second World War

In Padua, soon after Rossi's departure, almost all activities stopped due to the Second World War. At the end of the war, the situation of physics was a disaster, as in the whole of Italy. Scientists started coming back from the war, but they knew nothing about recent results and techniques. Moreover, funds for scientific activity were very scarce. Fortunately, the newly built Institute of Physics was not damaged. Antonio Rostagni,[17] who had been appointed professor of Experimental Physics in December 1938, decided after the war to complete Rossi's 1-million-volt accelerator. He thus founded a "Centre for the study of quick ions",[18] financed by the Italian National Research Council (CNR),[19] but

Berlin at Lise Meitner's laboratory. See: Alberto De Gregorio, *Via Panisperna e la radioattività indotta dai neutroni*, https://studylibit.com/doc/7563615/via-panisperna-e-la-radio attivit%C3%A0-indotta-dai-neutroni (accessed 15 June 2022); Edoardo Amaldi and Emilio Segrè, "Segno ed Energia Degli Elettroni Emessi da Elementi Attivati Con Neutroni", *Il Nuovo Cimento* 11 (1934), pp. 452–460.

16 Giovanni Polvani, "La Camera di Wilson dell'Istituto di Fisica della R. Università di Milano", *La ricerca scientifica* 12 (1941), pp. 410–420.

17 Antonio Rostagni (1903–1988) became professor of Experimental Physics at the University of Messina in 1935. He moved to Padua in 1938, where he remained until 1973. He co-founded the National Institute of Nuclear Physics (INFN), acting as the vice president of it from 1956 to 1958, and directed the Research Division of the U.N. International Atomic Energy Agency (IAEA) from 1958 to 1959.

18 "Centro per lo studio degli ioni veloci". See: Atti del CNR., *La Ricerca Scientifica* 17 (1947), pp. 1175–1177; Antonio Rostagni, "Centro di studio degli ioni veloci – Attività svolta durante il 1948", *La Ricerca Scientifica* 19 (1949), pp. 33–36.

19 This new research centre constituted an important step forward, which led Padua to play a major role at a national level. Padua was, for instance, to become, together with Rome

technical problems and the lack of funds forced him to give up in 1948, and he decided to focus on cosmic-ray researches. This provided a less expensive way to study particle physics, and similar choices were made in many European laboratories.

From 1948, Rostagni also started setting up contacts with foreign laboratories, inviting scientists to give talks in Padua and sending young Italian scientists to various European physics institutes, in particular in the United Kingdom. In February and March 1948, Rostagni himself visited physics laboratories in London, Bristol, Manchester, Birmingham, and Cambridge. His report on this trip shows the amazing gap in those years that separated the United Kingdom from Italy regarding the situation of physics funding, equipment, and researchers.[20]

The idea of having a cloud chamber is first mentioned by Rostagni in a 1949 report describing the activities carried out in 1948 by the Centre for the Study of Quick Ions.[21] Rostagni wrote that, in 1948, Padua physicists had started studies "for the construction of a Wilson chamber to be used with the large electromagnet. Dr. A. Loria has thus been sent to Manchester, at P. M. S. Blackett's laboratory, to learn about the chambers used there".[22] It is worth noting that in his reports on the activity of the Centre – Rostagni wrote such reports from 1949 to 1952 – he never mentioned Rossi, nor did he attribute to him any part of the available scientific equipment, be it the 1-million-volt generator or the

---

and Turin, one of the three first founders of the Istituto Nazionale di Fisica Nucleare (INFN) in 1951 (Milan joined right after). Edoardo Amaldi, "Gli anni della ricostruzione", *Giornale di Fisica* 20 (1979), pp. 186–225.

20 In all the places he visited in the United Kingdom, Rostagni observed an intense scientific activity related, "according to the English themselves", to "the notion abruptly acquired during the war" of the "capital importance of scientific research for technical progress and national defence" ("a detta degli Inglesi stessi [...] la nozione, bruscamente acquisita nel corso della Guerra [...] dell'importanza capitale della ricerca scientifica per il progresso tecnico e per la difesa nazionale"). See: Antonio Rostagni, "Laboratori di fisica in Inghilterra", *La ricerca scientifica* 18 (1948), pp. 529–535: 531. Rostagni underlined how the English administration helped physics laboratories: he was struck not only by the expensive equipment, but also by the number of people working in physics research and teaching. He pointed out that university assistants had a reasonable salary and physics students had good opportunities to be employed in universities or in industry, while in Italy university assistants were poorly paid, the profession of the physicist was almost unknown, only few industries had laboratories, and they often preferred to employ engineers.

21 Rostagni, "Centro di studio degli ioni veloci – Attività svolta durante il 1948", *op. cit.* (n. 18).

22 "Si sono infine iniziati gli studi per la costruzione di una camera di Wilson da usare col grande elettromagnete. Allo scopo il dr. Loria è stato inviato a Manchester, presso il prof. P. M. S. Blackett, a impratichirsi colle camere ivi in funzione", Rostagni, "Centro di studio degli ioni veloci – Attività svolta durante il 1948", *op. cit.* (n. 18), p. 34.

7-ton electromagnet.[23] Of course, these reports only provided short scientific news, but no mention of Rossi can be found in longer papers either.[24] We can wonder about the relationship between the two men – Rostagni took up Rossi's position only two months after Rossi was expelled – but these omissions may also be related to the fact that many of Rostagni's papers were intended for the CNR, which was financing the Padua research Centre, and Rostagni needed to highlight as much as possible the results he and his groups were achieving.[25]

The case of the cloud chamber is in this sense particularly interesting. In the report on the 1950 activities of the Centre, Rostagni wrote that one of the Padua research groups was working on the "construction of a Wilson chamber, to be used with the electromagnet that is now located at the Fedaia", and he specified that they expected it to "be put in use from next summer", i.e. in 1951.[26] As Rostagni specified, the cloud chamber group was made up at the time of "A. Loria, G. Zago and M. Cresti".[27] In his report for the activities carried out in 1951, Rostagni gave further details: "The Wilson cloud chamber built by Cresti, Loria, and Zago has started working between the pole pieces of the electromagnet at Marmolada. After preliminary tests, a study of the momentum spectrum of particles in extensive air showers at 2000 metres has started".[28] Rostagni never mentioned Rossi, and this seemed to be a totally new chamber.

However, the young physicists who worked with the cloud chamber after the Second World War, declared in recent years that the device they used, i.e. the one mentioned by Rostagni, was built by Rossi before the war. Marcello

---

23 Other reports by Rostagni: Antonio Rostagni, "Centro di studio degli ioni veloci – Attività svolta durante il 1949", *La ricerca scientifica*, 20 (1950), pp. 282–287; Rostagni, "Centro di studio degli ioni veloci – Attività svolta durante il 1950", *La ricerca scientifica* 21 (1951), pp. 1173–1184; Rostagni, "Centro di studio degli ioni veloci – Attività svolta durante il 1951", *La ricerca scientifica*, 22 (1952), pp. 911–921.

24 See for instance Antonio Rostagni, "Un osservatorio per raggi cosmici alla Marmolada", *L'energia elettrica* XXV (1951), pp. 211–215.

25 Rostagni actually started to explicitly mention Rossi, the Institute he built in Padua, and the fascist laws that led to his departure, in some of his general presentations from the 1960s.

26 "… alla costruzione di una camera di Wilson, da usare coll'elettromagnete che si trova attualmente al Fedaia. Si prevede che essa possa entrare in funzione nella prossima stagione estiva", from Rostagni, "Centro di studio degli ioni veloci – Attività svolta durante il 1950", *op. cit.* (n. 23), p. 1177.

27 Rostagni, "Centro di studio degli ioni veloci – Attività svolta durante il 1950", *op. cit.* (n. 23), p. 1177.

28 "La camera di Wilson costruita da Cresti, Loria e Zago ha cominciato a funzionare fra le espansioni polari dell'elettromagnete alla Marmolada. Ultimata la serie delle prove preliminari è stato ora iniziato uno studio dello spettro dei momenti delle particelle degli sciamo estesi a 2000 m", from Rostagni, "Centro di studio degli ioni veloci – Attività svolta durante il 1951", *op. cit.* (n. 23), p. 915.

Cresti (1928–2020), for instance, who graduated from Pisa in 1950 and immediately joined the cloud chamber group in Padua, said so in an interview recorded in 2015,[29] and so did Luciano Guerriero at a conference held in Padua in 2006.[30] A bit younger than Cresti, Guerriero (1930–) had started working with the cloud chamber on the occasion of his master's degree research work in 1951 – Guerriero then graduated in 1952. As for Guido Zago (1923–2022), who had graduated from Padua a bit earlier, in 1948, he confirmed the attribution of the cloud chamber to Rossi in private conversations with his friend Gianantonio Salandin, the physicist who set up the Padua History of Physics Museum in 1995.[31] Moreover, no controversies ever emerged about this attribution in recent years at the University of Padua: everybody agreed that the cloud chamber had originally been made by Bruno Rossi. Rossi himself recognised it as his own device when he came to Padua in 1987 and personally told this to Salandin.[32]

This raises quite an interesting issue about primary sources, as we have here discrepancies between Rostagni's written reports and several other accounts. Some of these accounts were recorded or written, as in the case of Cresti's and Guerriero's evidence; but others were only orally transmitted, as in the case of Zago's and Salandin's accounts, which we and other colleagues directly heard from Salandin. We can wonder about the value of accounts that left no tangible trace, except in our own memories. In any case, it is here quite clear that Cresti, Guerriero, and Zago had no reason to lie when they attributed the cloud chamber to Rossi,[33] while Rostagni had very good reasons to emphasise as much as possible the achievements of his group in his reports, as these were crucial to keep Padua's activities financed by the CNR.

29 Sofia Talas, *Intervista a Marcello Cresti, luglio 2015*, https://www.youtube.com/watch?v=LgoGSHs9_PI (accessed 21 February 2022).

30 Luciano Guerriero, "La capanna dei Fisici al Fedaia: un laboratorio internazionale per le ricerche sui raggi cosmici", *Proceedings of the Conference "La rinascita della fisica in Italia del secondo dopoguerra ai primi anni 1960: protagonisti a confronto", held in Padua, September 2006*. The proceedings of the Conference were not published but are kept at the Archives of the Department of Physics and Astronomy, University of Padua.

31 Gianantonio Salandin (1932–2018), worked on experimental high-energy physics, spending several years at MIT in Boston and at Brookhaven National Laboratory. Back to Europe in 1963, he collaborated with CERN in Geneva and became professor of Experimental Physics at the University of Padua. From the 1980s, he started taking care of Padua's physics heritage.

32 Salandin told us then about the history of the cloud chamber, when we started working in Padua.

33 They all worked with Rostagni for years and were on very good terms with him. When they were asked, they simply told the story in a very relaxed way.

Going back to the cloud chamber, we already said that we do not have any pre-war papers about it, so that it is difficult to know what the first version looked like. According to Cresti, the original cloud chamber had problems, and it "needed a lot of work, because it leaked; the block of bronze, or brass, on which the glass was fixed, was porous, and it was necessary to remake it with a less porous brass, so that it could keep the pressure".[34] However, it seems that no major changes were introduced to the original working mechanism in itself. Cresti explained that Guido Zago was in charge of the renovation project and made the chamber work. As for Cresti, he had just graduated and only helped Zago.

The first written description of the cloud chamber is provided in a paper published by Cresti, Loria, and Zago in 1953 (*Fig. 12.4*).[35] The device is based on the model described by C. T. R. Wilson in 1933: it has a fixed porous floor, which allowed a portion of the gas to escape when an elastic membrane moved.[36] The motion of the membrane, i.e. the expansion of the chamber, was achieved thanks to an electromagnetic mechanism, now missing. All the operating phases of the chamber were controlled by a circuit made of thyratrons, valves, and relays. Two Geiger-Müller counters were used to trigger the circuit. The chamber has an internal diameter of 25 cm and a depth of 10 cm. It is inserted in a copper cylinder that has, on its internal walls a coil made of a copper tube, where cold water circulated to keep the chamber's temperature controlled. The stereoscopic photographic camera, now missing, was designed and made by Zago. It could be inserted in the 10-cm diameter cylindrical hole in the pole piece of the electromagnet. It had two objectives and the advancement of the film was controlled by a small motor. A specially designed projector with two objectives was used for the spatial reconstruction of the particle trajectories and, to measure the curvature of the tracks, the group used a travelling microscope to obtain a plot of the particle's trajectory,[37] or a prism, so that the track curvature was compensated by a known superimposed curvature, in order to obtain a straight line.[38]

34 "... aveva bisogno di tanto lavoro perché perdeva, c'era il blocco di bronzo o ottone su cui poggiava il vetro era poroso ed è stato dovuto rifarlo, con un ottone meno poroso, in modo che tenesse la pressione", in Talas, *op. cit.* (n. 29).

35 Marcello Cresti, Arturo Loria, and Guido Zago, "Camera di Wilson in campo magnetico", *Il Nuovo Cimento* 10 (1953), pp. 843–850.

36 Wilson, *op. cit.* (n. 11).

37 Method used by Anderson in 1933; see: Carl D. Anderson, "Cosmic-Ray Positive and Negative Electrons", *Physical Review* 44 (1933), pp. 406–416.

38 The analysis then consisted in observing and compensating the deviations from straightness; see: Patrick M. S. Blackett, "Further Measurements of the Cosmic-Ray Energy Spectrum", *Proceedings of the Royal Society* 159 (1937), pp. 1–18.

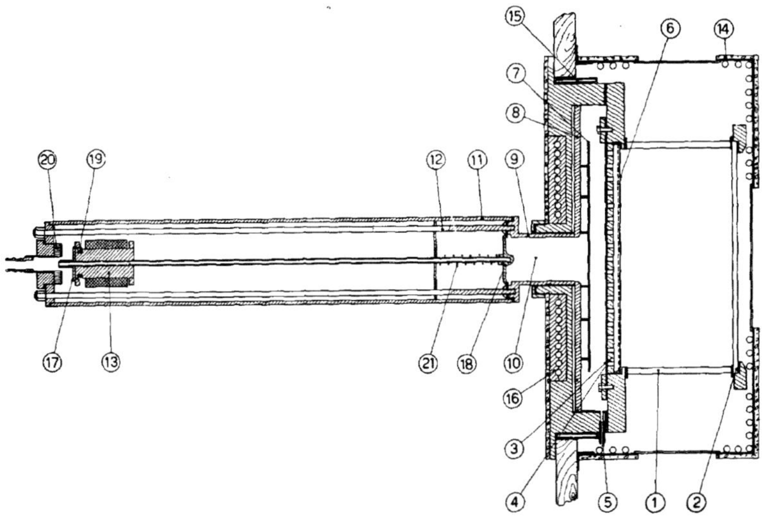

FIGURE 12.4 The cloud chamber described in 1953 by Cresti, Loria, and Zago, as depicted in Marcello Cresti, Arturo Loria, and Guido Zago, "Camera di Wilson in campo magnetico", *Il Nuovo Cimento* 10 (1953), pp. 843–850: 844

Let us add that Loria, one of the authors of the paper, was mentioned in Rostagni's reports as part of the early cloud chamber group. Born in 1914, he graduated in Padua in 1935 and was sent by Rostagni to Blackett's laboratory in Manchester several times and for various months. He was not directly involved in the renovation of the cloud chamber, because he was in Manchester at the time, nor did he spend long periods of time directly working with the cloud chamber, but he collaborated with the analysis of the photographs in Padua and co-authored all the papers related to the cloud chamber in those years. Older than the other physicists of the group, he was regarded as an expert in the field due to the experience gained in Blackett's laboratory.

As soon as it was ready, in mid-August 1951, the renovated cloud chamber was brought to a laboratory at Pian di Fedaia, close to the Marmolada, at an altitude of 2000 m in the Dolomites (*Fig. 12.5*). Rostagni had inaugurated this small laboratory the year before, in 1950, to study cosmic rays.[39] It was financed by the Società Adriatica di Elettricità (SADE), the private electricity society active in the Veneto region, which was at the time constructing a dam in that location. Rossi's 7-ton electromagnet was brought there at the end of July 1950 (*Fig. 12.6*). A group of physicists from Padua, guided by Piero Bassi, had started using the place to carry out observations by means of electronic

39 Rostagni, "Un osservatorio", *op. cit.* (n. 24).

FIGURE 12.5 The cosmic-ray laboratory set up in 1950 by Padua Institute of Physics at Pian di Fedaia, close to the Marmolada, at an altitude of 2000 m in the Dolomites
PHOTO COURTESY OF THE GIOVANNI POLENI MUSEUM, UNIVERSITY OF PADUA

instruments. In his 2015 interview, Cresti specified that he and Zago arrived at the Marmolada Laboratory on 15 or 16 August 1951, with the cloud chamber. At the end of the year, while they were setting up the apparatus, they were joined by Luciano Guerriero, who was working on his master's degree thesis, focused on the cloud chamber. After graduating in 1952, Guerriero went on collaborating with Cresti and Zago at the Marmolada Laboratory, and the three young physicists worked there for several years, until mid-1955 (*Fig. 12.7*). Sometimes they were all together but other times, for months, only one or two of them stayed at the laboratory, spending months alone.

They carried out experiments on several aspects of cosmic rays. They examined, for instance, the angular distribution of particles within extensive air showers,[40] and studied aspects of the so-called V particles that had been dis-

40 Marcello Cresti, Arturo Loria, and Guido Zago, "Sulla distribuzione zenitale degli sciami estesi", *Il Nuovo Cimento* 10 (1953), pp. 779–783.

FIGURE 12.6 Rossi's 7-ton electromagnet in use at the Fedaia cosmic-ray laboratory, in the 1950s
PHOTO COURTESY OF THE GIOVANNI POLENI MUSEUM, UNIVERSITY OF PADUA

FIGURE 12.7 Marcello Cresti (*on the left*), and Guido Zago (*on the right*) with two tourists, at the entrance of the Fedaia cosmic-ray laboratory
PHOTO COURTESY OF MARCELLO CRESTI'S FAMILY

covered in 1947.[41] Some experiments were performed in collaboration with Martin Deutschmann (1917–2008), one of Werner Heisenberg's collaborators

41 These new unstable particles were called "V particles", because of the shape of the observed events. They were widely studied at that time. See: George D. Rochester and Clifford Butler, "Evidence for the Existence of New Unstable Elementary Particles", *Nature* 160

from the Göttingen Max-Planck-Institut.[42] Deutschmann was particularly interested in the study of the collisions between pions and lead nuclei.[43] He brought to the Italian laboratory a multiplate cloud chamber from Göttingen, which was used together with Rossi's chamber in the magnetic field produced by the 7-ton electromagnet.

In the summer of 1953, a group of physicists from University College London (UCL) and Edinburgh University also started working at the Marmolada Laboratory, with their own cloud chamber, a high pressure one. Guerriero underlined that Harrie S. W. Massey, from UCL, had visited several European laboratories and finally asked to be hosted at the Fedaia Laboratory. According to Guerriero, for periods of time, the place, usually so quiet, looked like "an international laboratory", with nice opportunities for discussing and sharing knowledge.[44] Jim Grozier, in his account on the history of high energy physics research at UCL, explains that their cloud chamber, nicknamed "Susie", operated for nearly three years at the Fedaia Laboratory. He underlines that the building of this cloud chamber has been described as "the origin of HEP research at UCL".[45]

3 Interpreting the Padua Cloud Chamber Today

Cresti underlined that he and his colleagues carefully carried out their experiments, but only few relevant results were achieved at the Fedaia Laboratory in those years. According to Cresti, this was due to two main reasons. Firstly, the altitude of the laboratory, at 2000 metres, was too low compared to the other laboratories active in those years, at more than 3000 metres above sea

(1947), pp. 855–857; Charles Peyrou, "The Role of Cosmic Rays in the Development of Particle Physics", *Journal de Physique*, C-8 (1982), pp. 7–66.

42 Martin Deutschmann, Marcello Cresti, W. D. B. Greening, Luciano Guerriero, Arturo Loria, and Guido Zago, "An Anomalous V_0 Event", *Il Nuovo Cimento* 3 (1956), pp. 180–183.

43 Marcello Cresti, W. D. B. Greening, Luciano Guerriero, Arturo Loria, Guido Zago, and Martin Deutschmann, "Inelasticity in Collisions Between Pions and Lead Nuclei", *Il Nuovo Cimento* 4 (1956), pp. 747–757.

44 Guerriero, *op. cit.* (n. 30). The UK group of physicists published several papers about their experiments at the Fedaia Laboratory. See for instance: Joint Group from University College, London, and the University of Edinburgh, "The High-Pressure Cloud-Chamber Cosmic-Ray Programme at La Marmolada", *Nature* 175 (1955), pp. 445–448; Eric H. S. Burhop, "Cloud Chamber and Associated Techniques", *Nature* 175 (1955), p. 832.

45 Jim Grozier, A History of High Energy Physics Research at UCL, https://www.hep.ucl.ac.uk/history/history_cloudchambers.shtml (accessed June 2022) and https://www.hep.ucl.ac.uk/history/webArchive2013/history_old.shtml (accessed March 2022).

level. As for the cloud chamber, Cresti said that "it was a cutting-edge device at the time Rossi made it, but 15–16 years were elapsed, so that it was no longer cutting-edge and it was small, of a diameter of 25 cm, insufficient in comparison to the devices of about one square metre" that were in use in the 1950s.[46]

At the time it was used, several years after its construction, the Padua cloud chamber was thus obsolete and brought only poor scientific results. As a scientific instrument intended for research, it can today be regarded as a failure.

However, both Cresti and Guerriero pointed out how important the period spent at the Marmolada Laboratory was for the young scientists who worked there.[47] They had to take care of every aspect of the work: they not only had to keep the cloud chamber working, but also the illumination system and the stereoscopic camera. Furthermore, they carried out the first analysis of the photos they obtained, then the more accurate studies were performed in Padua, at the Institute of Physics. The young physicists thus learned a lot, partly thanks to Guido Zago, as Cresti underlined, and partly by working by themselves, having to study and handle all aspects of cloud chambers techniques. In fact, when they left the laboratory in 1955, they were ready to carry out research at an international level.

Cresti, for instance, after leaving the Fedaia Laboratory, moved to Göttingen for a couple of months to work with Deutschmann on the analysis of the cloud chambers photographs obtained at the Fedaia laboratory. He then worked in Padua on nuclear emulsions, introducing in this field the knowledge he had acquired in the Dolomites on the techniques for 3D–spatial reconstruction of tracks. A year later, he moved to Berkeley, at the University of California Radiation Laboratory, where he worked with Luis Alvarez's bubble-chamber group for two years. In Berkeley, his experience in analysing cloud chamber pictures was very useful in the study of bubble chamber pictures. On his second year there, he officially became a member of the laboratory staff. His experience in the US then led him to give interesting contributions when he got back to Padua, in the late 1950s. He launched, for instance, the construction of a Frankenstein – a semi-automatic device for the analysis of the output of bubble chambers – which was to be the first in Europe, and built an electrostatic beam separator, which was to be widely used at CERN for several years. Cresti actually became a leading figure in CERN.[48]

As for Guido Zago and Arturo Loria, they collaborated with the design and construction of the 2-litre propane bubble chamber that was made in Padua in

46 Talas, *op. cit.* (n. 29).

47 Guerriero, *op. cit.* (n. 30); Talas, *op. cit.* (n. 29).

48 See for instance: Alessandro Bettini, *Marcello Cresti (1928–2020)*, https://www.sif.it/riviste/sif/sag/ricordo/cresti (accessed 15 June 2022).

1956.[49] This was the first bubble chamber used at CERN.[50] Zago and Loria's pioneering experience in illumination issues and track reconstruction was highly regarded at an international level in those years.[51] Finally, Luciano Guerriero also had a brilliant career: after his activity at the Marmolada Laboratory, he worked on nuclear emulsions exposed to accelerator beams, collaborating for instance with Berkeley's Bevatron, then focused on space issues from 1978. He was the first president of the Italian Space Agency from 1988 to 1993.[52]

As we just saw, for Guerriero, as well as for Cresti, Zago, and Loria, the years spent working with the Padua cloud chamber marked his career. This actually sheds light on possible new ways to regard the cloud chamber itself.

The latter was certainly quite a failure as a scientific research instrument, but its biography is almost paradigmatic of the history of Italian physics right before and after the Second World War: while the 1930s were marked in Italy by the presence of scientists like Bruno Rossi and Enrico Fermi, the post-war period was seen by Italian physicists as a disaster, because knowledge, funds and instruments were missing and everything needed to be "reconstructed". The rebirth of Italian physics then took place in a period of about ten years, from 1945 to 1955, during the golden age of cosmic ray-based research. The Padua cloud chamber is material evidence of both the pre-war brilliant period and the post-war disaster and "reconstruction" of Italian physics. This could be a possible way to interpret and display the cloud chamber in a museum today, underlining how its biography is tightly related not only to the scientific issues of those years, but also to the political and economic situation.

Furthermore, being related to the name of Bruno Rossi, one of the most well-known physicists of the twentieth century, the cloud chamber could simply be seen as an icon of his presence in Padua. However, it is worth going further: we saw that this device was an essential element in the early research activities of a group of young physicists who, thanks to this device, learned a lot on experimental physics research. We suggest that the Padua cloud chamber could thus also be seen as a "training in research" instrument, going beyond the traditional categories of research or teaching scientific instruments.

---

49 Arturo Loria, Paolo Mittner, Igino Scotoni, and Guido Zago, "Camera a bolle a propano", *Il Nuovo Cimento* 11 (1959), pp. 718–724.

50 Arturo Loria, Paolo Mittner, Renato Santangelo, Igino Scotoni, and Guido Zago, "The Scattering of Positive 120 Mev Pions on Protons", *Il Nuovo Cimento* 22 (1961), pp. 820–843.

51 They were invited as experts in CERN meetings on bubble chambers. See, for instance, the proceedings of the "Informal meeting on heavy liquid bubble chambers", held at CERN, Geneva on 26 and 27 January 1959: CERN 59–24 / Proton Synchrotron Division, *Informal Meeting on Heavy Liquid Bubble Chambers*, 2 July 1959, CERN, Geneva.

52 For a short biography of Luciano Guerriero, see http://www.gapsrl.eu/gapsito/GAP200709/luciano_guerriero.html (accessed 15 June 2022).

CHAPTER 13

Failures, an Instrument Maker's Perspective

Nicolàs de Hilster

Although I did not attend the Scientific Instrument Commission's meeting in 2019 in Cuba, I noticed that during the earlier session on failures held in the Netherlands, failures were described as a function of the usefulness of the final instrument. As a very nice example, Richard Dunn opened the 2018 failures-session with a marine compass patented by Henry Jennings in 1818 (*Fig. 13.1*), which was designed to counteract magnetic flux from sources other than the Earth's magnetic field. Because it failed to do so, the compass was designated a failed instrument.

In this paper I would like to look at so called "failed instruments" in a very different way and challenge their ill-fated status. Is an instrument that failed to produce an anticipated result a failed instrument? Do failed instruments

FIGURE 13.1 Jennings insulated compass
NATIONAL MARITIME MUSEUM, GREENWICH, ACO1517.
PHOTO © NATIONAL MARITIME MUSEUM, GREENWICH

 DOI:10.1163/9789004689107_015

actually exist? Being an instrument maker myself I would like to approach the discussion from an instrument maker's point of view and show that there is no such thing as a failed instrument.

When we look at the process that results in the making of an instrument, several stages can be discerned (*Fig. 13.2*). First of all, we have the *issue stage*, in which an issue is addressed that we would like to solve using an instrument or an improved version of it. This stage is followed by a *philosophical stage* in which various options to deal with the issue are concocted, rejected, and improved upon. Once that stage offers a clear outline of the instrument that is expected to solve the matter, the final design is materialised by an instrument maker in the *creative stage*. This instrument maker can be the very same person who went through all of above stages, or a person who is commissioned to create the instrument. In the latter case, an additional *documenting stage* is required in which the *philosophical stage* is transferred to paper, either in the form of a drawing or as a detailed description, or both. It has to be noted that each part of this process can be done by separate individuals. So, a mariner might be very unhappy with his ill-functioning compass and ventilate about it to the manufacturer, who in turn could think of new ways to produce a better version of it and instigate one of his draughtsmen to make drawings, which would then be given to the instrument maker(s) in the manufacturer's workshop to create the newer instrument.

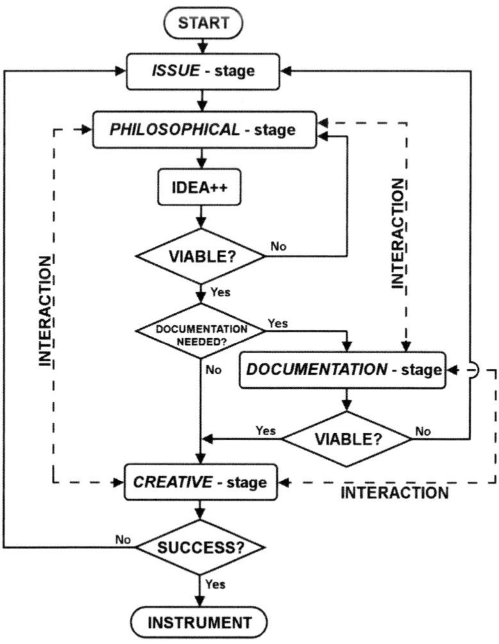

FIGURE 13.2
Flow diagram of the creative process
DIAGRAM BY THE AUTHOR

Now it may not be obvious from the above story, but during this process, multiple instruments would be created, no matter whether each stage is done by another individual or not. Even though the outcome of the process is a single physical instrument, each stage, apart perhaps from the *issue stage*, has its own output of hypothetical instruments (the *issue stage* is where an instrument already exists or a new one is required to solve a problem). The most fruitful of these stages is the *philosophical stage* where perhaps a dozen or more instruments are imagined, but of which only one (or a few at the most) will find its (their) way to the *documenting stage* or directly to the *creative stage*. While documenting, the instrument is again built mentally to ensure that the documentation will result in an actual instrument and does not present the instrument maker with new challenges. In case the documentation results in insurmountable problems, the design has to go back to the *issue stage* and *philosophical stage* in order to be improved.

So by the time an instrument arrives at the *creative stage* it has been mentally created and recreated many times. If this mental process was done by multiple persons, there is a fair chance that this process was a bidirectional one, with interaction between various stages. But even when done by the instrument maker alone, there is a fair chance that he expressed his ideas to others to test his or her ideas. Each stage also creates a different kind of instrument. The *philosophical stage* creates hypothetical instruments that solves the issue, but may not yet result in a feasible construction. The *documenting stage* creates a description and, when drawings are made, a feasible construction plan. Finally in the *creative stage*, the instrument maker will try to create the physical instrument. Of course, there may be further interactions at this stage, and we are all too familiar with the expression "going back to the drawing board". Only if the instrument maker, or the commissioning party, thinks that what was created in the *creative stage* was successful will the process end with a physical instrument.

In the past two decades I have been through all these stages many times. Apart from the nautical replicas and reconstructions I have made, I have been restoring some of the geodetic instruments in my collection and created several geodetic tools and targets for my business. In a few cases that work has been done with others.

For example, for a mariner's astrolabe, dimensions of the original were taken by the then curator at the National Maritime Museum, Richard Dunn; drawings were made by myself; the initial construction built by Günther Oestmann; and the final adjustments and calibration performed by myself again. In this case the *philosophical stage* for this type of mariner's astrolabe had been done

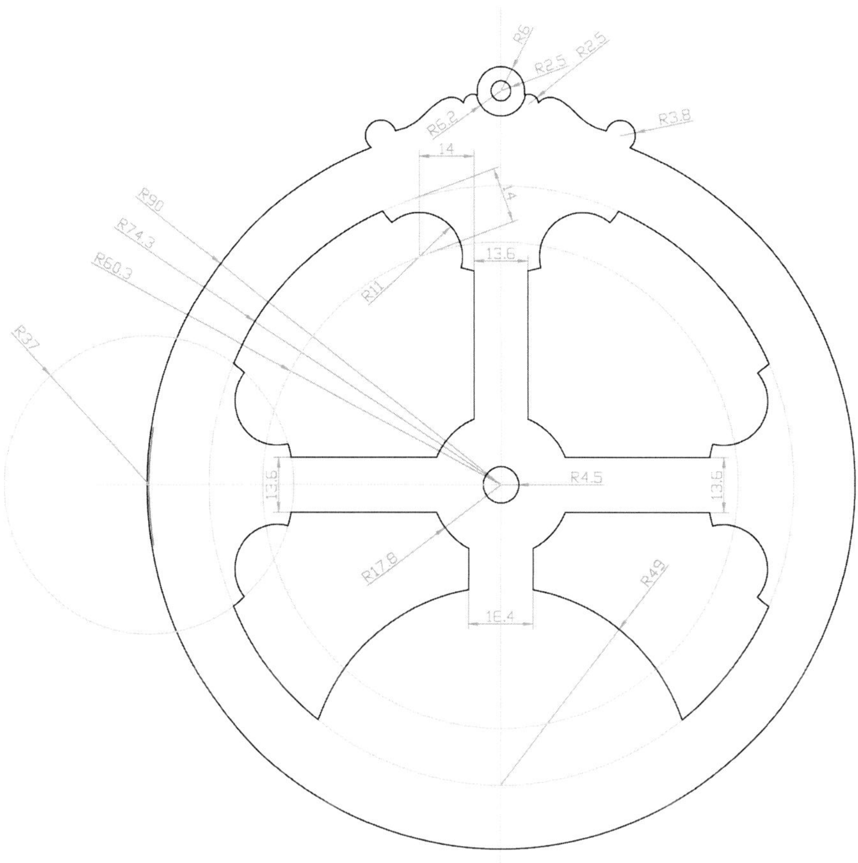

FIGURE 13.3 The mariner's astrolabe as a function of circles
DIAGRAM BY THE AUTHOR

in the sixteenth century, but I went through that stage myself to see if I could explain the design features by using a compass and ruler only (*Fig. 13.3*).

Another example of collaboration was the creation of an optical tooling target created for the accurate alignment of ship gyros (*Fig. 13.4*). Here the design was made on-the-fly and solely based on the parts that had to be included: the prism, the circular bubble level (also known as a circular vial), and the coincidence level (also known as a coincidence vial or split-bubble level). I was assisted by an apprentice instrument maker, who took care of milling the base plates, designing the counteracting springs, and creating the first version of the prism-holder. The development of this instrument went through the *issue, philosophical,* and *creative stages*. The *documenting stage* was skipped and

FIGURE 13.4 Optical tooling target from the author's workshop
IMAGE BY THE AUTHOR

replaced with what could be called the *interaction stage* in which my assistant and I discussed various details, alternatives, and problems to overcome.

The reason why I show these two examples is that the creation of both instruments did not go smoothly, and both could have ended up becoming what is called here a failed instrument. In the case of the astrolabe, the pointers of the alidade were not properly shaped. The fiducial edges were made perpendicular instead of bevelled, making the edge hard to read. Moreover, the two pointers were not in line with each other (i.e., their fiducial edges not encompassing 180°), so that readings would be produced with errors well in excess of the instrument's theoretical accuracy of about 10 arcminutes. In addition, the pinnules were given a smooth finish using a countersink, which created observational errors as well.[1] In the case of the optical tooling target, the

1 Nicolàs de Hilster, "Observational Methods and Procedures for the Mariner's Astrolabe", *Mariner's Mirror* 100, no. 3 (August 2014), pp. 261–281: 269–270.

first prism holder was not made accurate enough to allow it to be adjusted down to its required accuracy of 1 arcsecond, while the front and back of the instrument, which should have been parallel within that value, deviated from each other by approximately 14 arcseconds. In addition, the instrument was finished the day before it had to fly to the field location in the Middle East, but the calibration method to properly align the prism with the base and with the coincidence level had yet to be devised.

So, the astrolabe was dismantled and reworked, straightening the pointers, giving them a bevelled edge as was usually done with period astrolabes, and removing alignment errors and countersunk edges of the pinnules. At the same time the surface and suspension ring, which still showed lathe marks, were smoothed and given a new patina. For the optical tooling target a new prism holder was already made during the *construction stage*, but the calibration procedure was created the night before departure. On site the instrument was calibrated twice to see if it gave repeatable results, which it luckily did. Both instruments now work flawlessly (provided they are properly used), although the optical tooling target still has a misaligned rear edge, but as the deviation is known, this is not a real issue anymore, and it has been successfully employed in the field (*Fig. 13.5*).

Given these two examples I think we may need to redefine the term failed instrument. Although the above cases show that things can go wrong in the creative process, they also show that an instrument, once it has left the workshop, normally is what it is supposed to be. The above two cases show that it may not yet be perfect, but the issues in both instruments have been addressed and corrected. This means that there was a temporary failure at the most.

So this brings us to the question of what an instrument is and whether or not it can be a failure at all? To me as an instrument maker, an instrument is a contraption that has been made according to a certain philosophy and perhaps according to a certain description. As such, when an instrument leaves the workshop other than in the bin, it cannot possibly fail, unless it was not built according to specifications or was not finished. In principle, each instrument leaving an instrument maker's workshop is a finished product, so how can it be a failure?

We can approach the problem the other way around. So, is an instrument that does not create a successful observation or measurement a failed instrument? If so, what about all the prototypes that have been built, but required further adjustment? Are these all failures? If so, then we have failures in abundance! I think, however, that such an instrument would go back into the process, and being the issue, it would most likely go all the way back to the

FIGURE 13.5 The optical tooling target in action
IMAGE BY THE AUTHOR

philosophical stage, as some proper thought is needed to find out why it did not work and what should be done to correct it.

A good example of this is the Hubble Space Telescope, which was launched in April 1990. Soon after it launched it was realised that the telescope did not perform as expected. It should have been able to meet a Rayleigh criterion (i.e., image resolution of two objects) of at least 0.1 arcsecond, but instead it produced images at a 0.7 arcsecond resolution, so seven times worse.[2] The telescope's main mirror had been optically tested, using a 58-cm rod in the process. A centring cap at the end of the rod had an opening through which the highly polished end of the rod was visible (*Fig. 13.6*). Distance was measured towards this end for placing optical test elements using interferometry, but as a small section of anti-reflective coating was missing around the opening in the cap, the measurement was made at the cap instead of at the rod below it, creating

2 Lew Allen et al., *The Hubble Space Telescope: Optical Systems Failure Report*, NASA, November 1990, p. 5-1, on the NTRS – NASA Technical Reports Server, https://ntrs.nasa.gov/citations/19910003124 (accessed 22 February 2020).

a 1.3 mm error. Luckily the Hubble Space Telescope was launched in the Space Shuttle era, and the problem was solved by replacing the camera and installing a corrective optics package during a servicing mission in December 1993 (*Fig. 13.7*).[3]

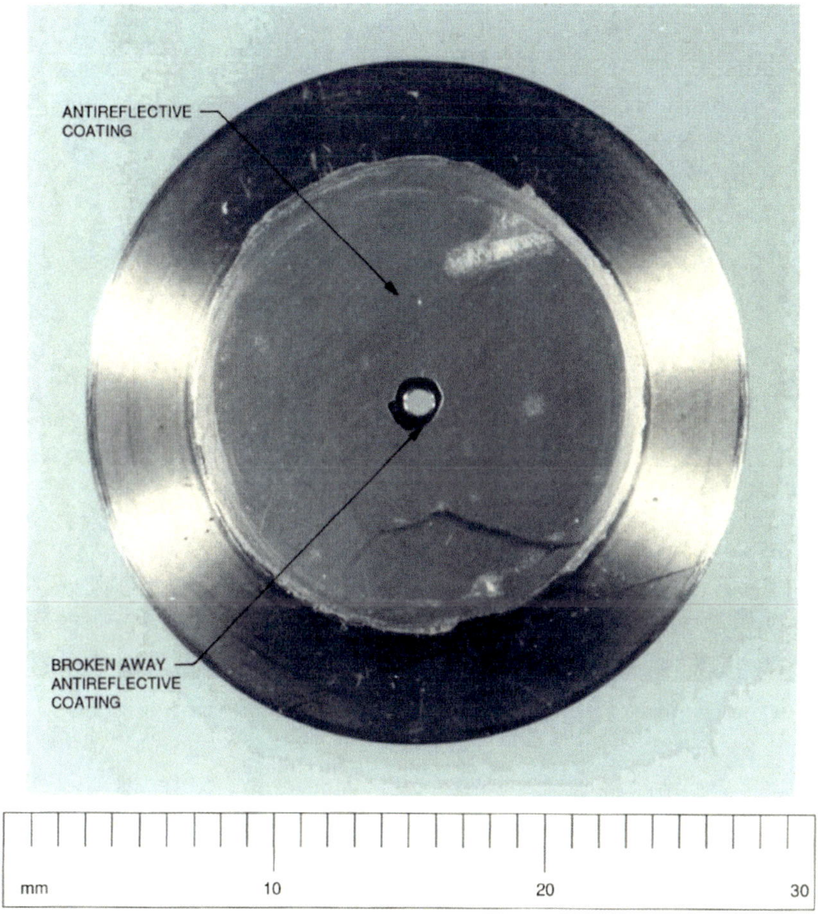

FIGURE 13.6 Top view of the field cap of the Hubble Space Telescope mirror testing interferometer. NASA, November 1990, from Lew Allen *et al.*, *The Hubble Space Telescope: Optical Systems Failure Report*, NASA, November 1990, fig. 7-5

3 NASA, image of servicing the Hubble Space Telescope, 5 December 1993, STS061-87-062, https://spaceflight.nasa.gov/gallery/images/shuttle/sts-61/html/sts061-87-062.html (accessed 17 August 2020, site discontinued).

FIGURE 13.7 Repairing the Hubble Space Telescope. NASA, 5 December 1993, STS061-87-062, from https://spaceflight.nasa.gov/gallery/images/shuttle/sts-61/html/sts061-87-062.html (site discontinued)

So was the Hubble Space telescope a failure, or was what we have seen just another stage in the process? Initially the Hubble Space Telescope did not live up to its expectations, but it went back into the process, and after the repair, it performed as intended, making the next version of it a great success.

At the moment that an instrument does not perform as expected, there are several options. It could well be that, for instance due to a lack of funding, the development process stops, and the instrument never matures. Or it could return to the *philosophical stage*, but a solution to the issue simply is not found. Or perhaps it did return, and a theoretical solution was found, but the improved instrument could not be created due to technical limitations. Of course, we hope that it does go through all of these stages, just like the Hubble Space Telescope did, resolving the issue in the process.

But what if going back into the process does not solve the issue, what exactly is it that has failed? Was it the instrument, the designer, the instrument maker, or the state of technology that has failed? Stating that the instrument failed sounds to me like blaming the instrument for something it cannot be held responsible for and making it the scapegoat of our own human failure.

General Index

Academy of Lausanne 21
ACO storage ring, Paris 141
Adams, J. C. 69
Adams, John 93
Admiralty, the British 6, 7, 14, 16, 17, 96, 97, 98–100, 103, 104, 105, 107, 114, 117, 123, 200
Admiralty Compass Committee 6, 7, 12, 13, 14
Admiralty Compass Department 6
Admiralty Compass Observatory 1, 6–20
 catalogue 7–11
 Deptford Dockyard 7
 Dutton Park, near Slough 7
 museum 13, 14, 17, 19, 20
 Woolwich 7
aircraft 106
air-moisture content, measurement *see* humidity
Alexander, [George] 16
Al-Karakī 148, 149, 150, 152, 166
Allerthorpe, Yorkshire 58
altazimuth mounts 50
altitude of the Sun 153
A.M. Nomikos Maritime Company 72
amateur astronomy in nineteenth-century Britain 57
AND logic gate 212
angular scales 4, 167–177
 division and reading 167
Apelquist, Carl 41, 42, 43, 47, 48–49, 52
 jackscrews 42
 machine tools 42
Aristotelian physics 172
arithmometer 195
armillary sphere 44
Armstrong, Sir W. G. 60
Army, British 20
asteroid 9 Metis 69
astrolabes 150, 153, 154, 155, 157
astrology 80
astronomical observations 84, 88, 89, 117
 lunar 50, 51
 nebular 50, 51, 55
 research 141, 170
astronomers 4, 151
 Greek 2
astronomy, spherical 151, 164
astrophotography 69
Athens 71–75
atmosphere, state of 23
 atmospheric pressure 23, 31
 relative humidity 23, 28
 temperature 23
August, Ernst Ferdinand 25
Austin, John 129
Australia 202–208
authorities 4
azimuth 153

babies, flat headedness in 81
Baines, Robert Raines 100, 102, 103, 106, 120, 130
Baldini, Ugo 175
ballistic missile 138
Bamber, James 130
Banks, Joseph 179, 186
 Young's eriometer 179–181
barometer 24
Barlow, Peter 19
Barrow, [Henry] 9, 11
Bassett, John 128
Bassi, Piero 219
Beaufort, Sir Francis, Hydrographer to the Navy 6, 7, 19
Beaver Cay 206
Bedford, Walter 127
Belgium 23
Berlin 25, 27
 Berliner Elektronenspeicherring-Gesellschaft für Synchrotronstralung (BESSY II) 143
Bern 27, 32, 37
 Meteorological Observatory 21, 27
 Director 21, 27
Bethe, Hans 214
Bohr, Niels 214
Bill, Robert 113, 115, 116
 gas lighting scheme 115
 manufacturing pianofortes 115

Bingham, Arthur Batt 95, 99, 130
Binnig, Gerd 143
biological research 3
biologists 3
Birmingham 215
Blackett, Patrick M. S. 212, 214, 215, 219
Blunt, London 11
Blunt, New York 9
Board of Longitude 3, 14, 16, 17, 92, 93, 96, 98–99, 100, 102, 103, 114, 116, 117, 118, 119
Board of Trade 19
Bombay Engineers 7
Borgondio, Orazio 4, 167, 174–177
  Mémoires pour l'histoire des sciences et des beaux arts (1720) 174, 176
Boscovitch, Ruggero 175
Boswell, John Whitley 91, 129
Boston 109–110
  Boston Marine Society 109
Bouguer, Pierre 89–91, 122
  conical log and sinker 89, 91–92
Bourne, William 84
Bradford & Son, [George] London 8
Brahe, Tycho 167, 170–174
  Astronomiae instuauratae mechanica (1598) 170, 171, 172
brain, animal 81
  extinct 81
  shape 81
brain, human 77–83
  aging 78
  cognition 78
  shape 77
  size 77
  understanding of 78
brake-drum 58
Brazil 174
Bredman, Johannes 49
Bridges-Lee, John 195
Bristol 215
Brisbane Barrier Reef Committee 203
British Astronomical Association 60
Brombacher, W. G. 24
Brougham, Henry Peter 185
Brown, John 143
Browse, Henry 130
Brunner of Paris 194
Brunton, James, Compass Examiner 7–8
Bryden, D. J. 3
Bulgaria 63
Buonanni, Filippo 175
Burney, James 99–100, 130
Burney, William 88
Burtsal, William 126
Burt's buoy and knipper 111, 123

calculators, input-output 4, 150
Calais 58
Cambridge, UK 2, 57, 69–71, 215
Campbell, W. W. 70
Canadian Rockies 195
candle flame 183, 188
Canterbury 104
Capacci, Domenico 174, 175, 176
Capella 70
carbon 16
Carbone, Giovanni Battista 174, 175, 176
Cary, William 104
cascade voltage multiplier 27
  Cockcroft-Walton 27
  Greinacher 27
Cassella, Louis Pascal 195
Castrioto, Felix Antonio 127
CCD chip 135, 136
celestial mechanics 38
celestial observations
  see astronomical observations
CERN 224, 225
Céspedes, Andrés Garcia de 167, 172
  Regimiento de Navegación (1606) 172
Ceylon 204
Chance Brothers, Birmingham 59
Chicago 61, 214
Christie, Samuel Hunter 7, 12
Christie, W. H. 69
chronometers 125
Clark, Alvan 61
Clark, Alvan and Sons 61
Clavius, Christopher 167, 173
Clay, baked 21
Clement, Leandre 95, 103–104, 131
  current meter 103
  engine indicator 103–104
  sillometer 103–104
  submarine thermometer 103–104
cloud chamber, made in Padua 4, 209–225

Cockcroft, John 27
Cockcroft-Walton cascade voltage multiplier 27
Cockcroft-Walton generator, 1-million-volt 210, 211, 214
collection, reference 12
Combe, George 78, 80
 The Constitution of Man, 1828 78
comet 68
 comet of 1577 170
commercial success 125
Common, A. Ainslie 69
compass
 Admiralty Standard 7, 13
 azimuth 11
 binnacle 9
 boat's 8–9, 11
 Chinese 8
 Danish 12, 13
 development 6
 hanging 8
 magnetic 6–20, 153, 155, 156, 159, 161, 163
 needles 12, 17
 night illumination of the card 124
 originality 12
 tank 20
 technical details of 12
compilers of dictionaries and encyclopaedias 90–95
Compton, Arthur 214
computed tomography 144
construction of knowledge 1
consumers 4
Cook, Captain James 107
Cooke, Frederick 59
Cooke, John 106, 128
Cooke, Thomas 2, 3, 58–61
 first equatorial telescope factory in Britain 59
 death, 1868 60
Cooke, Thomas, jr. 59
Cooper, William 127
Copenhagen 214
Copernicus, Nicolaus 44
Cornell University 214
cosmic-ray research 4, 210, 215, 219
Cosnahan, Mark 105, 130
 auxiliary paddle steamer *Victory* 105
cost of production 3
Costa, Francisco da 172
 Arte de Navegar (1596) 172
Coventry 115, 116
craftsmen, a skilled class of 156
cranial cavity 78
Cresti, Marcello 216, 217, 218, 220, 222, 223, 224, 225
Crichton, Michael 80
Cronstrand, Simon Anders 38, 44, 49, 52, 55
Crowe, [of] Faversham 10
curators 1

Danger, R., micrometer 72
Darwin, G. H. 69
Dashtakī 149, 152, 153, 155, 156
Davies, S. B. 109–110
Davy, Humphry 186, 191
dead reckoning 84, 85, 88, 89, 125
 mile to a degree 85
Dearborn Observatory 61
Debenham, Frank 202
De La Rue, Warren 63, 194
Deluc, Jean-André 24
Denmark 13, 58
Denning, W. F. 64
Dent, [Edward] 9
 Dent's Axis [compass] 9
Deptford 107
Derby, named gentleman of 98
Desaguliers, J. T. 97
desiccant 22, 28, 31
Dession, Joseph 85
Deutsches Elektronen-Synchrotron (DESY) 139, 140
Deutschmann, Martin 222, 223, 224
deviation angle 159
Deville, Édouard 195
Dewhirst, D. W. 59
Dickman, [John], Leith 10
diffraction grating 70, 137, 138
diffraction, optical 4, 134, 135
diffraction ring of transmitted light 184
diffusion
 through porous materials 22–23
diffusion hygrometer
 see hygrometer, diffusion
distance run 84

Dollond, London 38
  10-foot refractor 38
Dolomites 219
Dorpat refractor 46
Dover 58
Drigo, Angelo 213
Dublin 71
Dudley, Robert 173–174
  Dell'Arcano del Mare (1646–47) 173
Dufour, Henri 22
Dufour, Louis 21, 22, 28, 29
Dundee 57, 58
Dunn, Richard 1–2, 226, 228
Dunsterville, compass maker 16

Earth's axis 44, 48, 49
Earth's magnetic field 14, 226
East India Company 88, 106, 107, 118
  ship *Worcester* 93
Edinburgh 78, 91, 190
Edinburgh Phrenological Society 78
Edye, Lieutenant, RN 10
Efstathiadis, N. 72
Ekström, Daniel 41
  workshop 41
electrotechnology 27
Elswick-on-Tyne 60
Elvey, John 98, 104, 131
Encyclopaedia Britannica 91, 110
Encyclopaedia Metropolitana 91, 122
Engineering magazine 195
England 53, 92, 103, 108, 120
EPFL, École Polytechnique Fédérale de Lausanne 21
equatorial mount
  for Herschel telescope 38–56
  Nairne 45–48
eriometer, Young's 4, 178–192
  as a measuring device 186
  determining the size of wool fibres 178, 181, 184, 186–187
  examples 179
  use 178, 183–185
European Astronomical Union, General Assembly 1955 71
Evans, Captain [Frederick] 19
Evans, George 95, 106, 131
Everest, Mount 198

expeditions 193
  Arctic, 1839 6, 19, 120
  Great Barrier Reef, 1928 4, 193, 201, 202–206, 208
  polar, 1773 100
  polar, 1777 89
  Shaksgam Valley, 1926 201

factories 57
failures 1–5, 6–20, 21–37, 38–56, 57–76, 77–83, 84–131, 132–147, 148–166, 167–177, 178–192, 193–208, 209–225, 226–235
  commercial 21–37
  context of 1, 4, 5, 133, 165, 193
  economic causes 3, 4, 208, 225, 234
  expected part of scientific practice 1, 133, 226
  from air pollution 57, 64
  from complexity 177
  from lack of logistical support 4, 197, 207
  from lack of management 55, 64, 76
  from wrong users 186–190
  human 235
  importance of 83
  integral part of scientific practice 1, 226
  of new technology to replace old 117–125, 192
  political causes 4, 225
  reasons for failure 3, 64, 76, 146–147, 177, 192, 193, 208
  social factors 3, 4, 90, 133, 165, 192
  spatial factors 193
  technical 55, 90, 234
  temporal factors 193, 224
  what can be learned from 1–5
Falconer, William 88
  Dictionary of the Marine 88, 95
farmers 183
  wool fibre diameter 178, 183
fascist racial laws 4, 214
Fergusson, Edward 129
Fermi, Enrico 213, 225
Ferndene, Gateshead 58, 61–69
  portable observatory dome 63
Fisher, George 17
First World War 196, 198, 199, 201

Flamsteed, John 45
Flinders Island 203
Flinders, Matthew 19
Florence 173
Flums, Switzerland 196, 201
fog 2, 201, 209
folk astronomical practices 148, 150
Fourcade, Henry 196
Foxon, William 91–92, 100, 106, 107, 108, 118, 126
France 13
Fraser, William 42
Fraunhofer, Joseph 46
Fremantle, Western Australia 204
French government 194
French, [Mr] 10
Fresnel zone plates 132, 134, 135
  as lenses 137
"Froehn" problem 22
fusion process in the Sun's core 136

Gabor, Dennis 132, 141
Gall, Franz Joseph 77–78, 80
Gateshead 58, 61–69
gazetteer 159, 161, 162, 163, 164
Geiger-Müller counters 212, 218
Geminorum 38, 39
Geneva 24, 27
Gentleman's Magazine 123
geodetical research 52
George, Captain [Christopher] 9
George I, king of Great Britain 96
German Volkswagenstiftung 141
Germany 57
  Lower Saxony 142
  mines 57
Gernandt, Edvard 52
Gibraltar 204
Gilbert, [William], London 8, 11, 12
  compound needle 13
  magnetic compass 12
Gilbert and Wright, instrument makers to the East India Company 93
Gill, David 63
Gilmore, Joseph 92, 126
Glaisher, J. W. L. 69
glass disks 59
  floated in mercury 60
gnomon, folding 159
Gompertz, Lewis 123
Goold Island 205
Gordon, L. D. B. 57
Gosport, naval academy at 88
Göttingen 190, 223
  Max-Planck-Institut 223
  University Observatory 132, 136, 141
Gottlieb, Valentine 91, 92, 100, 120, 128
  patents for cranes and axle-trees 91
Gould, Chester 106, 108–111, 119, 128
  Gould's perpetual log 90, 108, 119
Gould, Horace 108–110
Gower, Richard Hall 93, 94, 118, 119, 128, 129
  experimental vessel *Transit* 93
  improved patent log
Gowland, [George] 11
Graham, Andrew 69
Grant, Alexander 93
Grant, Charles, vicomte de Vaux 91, 92, 95, 129
  Grant's hydroscope 91, 94
graph 31
grating monochromator 136
Graves, Thomas 99, 131
Greinacher, Heinrich 2, 21, 27–32
Greece 71–76
Greenwich Observatory 45
Grozier, Jim 223
Grubb, Sir Howard 70
Grubb Parsons and Co. Ltd., Sir Howard 72
Grubb Telescope Company 70
Guerimand, Jacques 119–120, 127
Guernsey 96, 97
Guerriero, Luciano 217, 220, 223, 224, 225
Guildford 201
  Hog's Back 201
Gurney, Hudson 192
Gustav VI Adolf, king of Sweden 53

Haenni 36–37
Hall, Asaph 61
Hallowell, Rear Admiral [Sir Benjamin] 16
Hällström, Carl Peter 38
Hamburg 139
Hamill, Hugh 91, 106, 129
Hamilton, Thomas 91, 99, 106, 128

Hanisch, Paul 186
reconstruction of Young's eriometer 186–188
Hardy, Arthur Sherburne 194
Harkness, William 61
Harris, John 85
Harris, Sir William Snow 9, 10
heat-related problems 22
heaving the log 84, 85
Hebling, Dr Robert 196
Heerbrugg 196
Heidenhain enterprise in Traunreut 137
Heisenberg, Werner 222
Hellström, Peter 43–44
Herschel, William 2, 3, 38, 39, 51
discoveries of nebulae 41
discovery of planet Uranus 54
mirror 39–42
mount 39, 42–52
optics 39
portrait by Lemuel Francis Abbot 54
7-foot reflector 38–56
Higginson, Francis 100, 101, 131
Hilster, Nicolàs de 5, 196
Himalayas 201
Hinks, Arthur R. 197, 198, 199, 200
historians of instruments 20, 159
historical scientific apparatus 1
history 77
history of science 77, 150
importance of failures 77, 78–80
Hodge, Arthur 130
Holiday, Henry 64–68
Holland 97
The Hague 97
holographic techniques 132, 137, 141
Hookey, James 100, 101, 131
silver medal for fish-shaped log 100
Hubble Space Telescope 232–235
Huggins, William 63
Hull-built whaler 100
Hull glue factory 100
humidity 23
Hunt, Frederick 13–14
Hutchison, D. A. 197, 199, 200
Hydrographic Office, London 8, 9, 11
hygrometers
dew point 24, 25–26
diffusion 2, 21–37
double diffusion 27–29
absolute humidity 25
relative humidity 25
electrical 21
electronic recording 21, 26
gravimetric 24
hair 2, 23, 24, 36, 37
accuracy low 24
calibration 24
minimal maintenance 24
simple to use 24
modern capacitive (TESTO 608–H2) 33, 37
new types 37
simplified diffusion, Greinacher's 29–31, 35
synthetic fibres 24
wet-and-dry bulb 23–24, 25
hygroscope 24

Iberia 172
Istituto Superiore di Sanità, Rome 210, 211
images of biological specimens 132, 144
did not convince biologists 132
impeller 118, 120
clogged with weed 120
India 7, 106
industries, English 4, 178–192
agricultural 4, 178–192
medical 4, 178–192
textile 4, 178–192
innate conservatism of users 3
Institute of Mechanical Engineers 58
instruction manual 33, 158
instrumentation, scientific 1, 13, 89, 150, 153, 155, 156, 161, 167, 176
astronomical 2–3, 38–56, 57–76, 167–177
European 165
for nuclear particle detection 27
incorporating photography 193–208
magnetic 2, 6–20
navigational 6, 167–177
instrument maker 5, 45, 161, 164, 227–231
apprentice 229–231
designer 5
fabricator 5
perspective 5, 226–235
stages of making an instrument 227–232
usefulness of failure 226

interferometry 232
Iran 4, 162
  Isfahan 159
  Islamic society in 155, 165
  Safavid 165
Iraq 201
  Mosul boundary project 201–202
iron filings 16, 17
Italian National Research Council 214
Italian physics 5, 209–225
Italian Space Agency 225

Jarvis, Captain Thomas Best 7
Jennings, Henry Constantine 6, 11, 14–19, 226
  insulating compass 6, 11, 14–19, 226
Jesuits 167, 174–177
Jewish origins 167
João III, king of Portugal 167
  Duarte, Dom 167
  Henrique, Dom 167
  Luís, Dom 167
Jobin, A.-G. Ynon, precision polarimeter 72
Johnson, Dale L. 35
Johnson, Commander Edward John 7, 17, 19
  Superintendent of the Compasses 7
Johnson, W. 109
Johnston, Stephen 195
Jones, Thomas, London 10
Jouveneau, A. 23, 29
Junk, Andreas 3

Karlskrona 42
Kärnfelt, Johan 2
Kater, Captain [Henry] 9
Keen, [Robert John] 11
Klein, Joannes 176, 177
Knox, Barnabas C. 110
Knudsen Law 33
Kodak 204, 206, 207
Koehler illumination 137
Kontopoulos, George 71
Kortright, Lieutenant [A.] 11
Kundt, August 22, 29
Kurz, Jacob 173
Kynvyn, James 173–174

laboratory 22
Lamb, James 131
laser light 132
  radiation 137
  red, green, blue lasers 137
  Spectra Physics laser 137
Lassell, William 61
latitude 14, 49, 89, 125, 151
Lausanne 22, 32, 35
  Academy 22
Laussedat, Aimé 194
Lavanha, João Baptista
  Tratado del Arte de Navegar (1588) 172
Lazos, Panagiotis 2
Lebanon 148
Leguin, Estiennne 91, 92, 98, 119, 127
lenses for a microscope 132, 135
Lick Observatory 70
Liddell, Charles 58
life sciences 133
light, particle nature of 186
Lindblad, Per Olof 53
Lindsay, Lord James 63
lists of *qibla* values 156
Liveing, G. D. 69
Liverpool 109, 113
  Liverpool Committee of the West Indies Association 113
  Mersey estuary 113
Lockyer, Sir Norman 70
log and line 84, 85, 89, 95, 106
  alternative log designs 84–131
  problems with 87–90
London 42, 96, 97, 104, 109, 113, 190, 215
  canal in St. James's Park 97
  Clerkenwell 116
  International Exhibition, 1862 59
  Middlesex Hospital 192
  nautical academy 122
  River Lea 93
  River Thames 96, 97
  The Medway 103–104
  Woolwich Dockyard 104
longitude 14, 89, 122, 125, 151
Loria, Arturo 215, 216, 219, 224, 225
Los Alamos 214
Lothian Birth Cohort project 78
Loude, Jean-François 2
Lovelace, William 107
Lück, Winfried 35
lunar-crater sketch, first 65, 67

lunar occultation 38
Lusverg, Dominico 174

MacArthur, J. J. 195
McAuley, [of] Dublin 9
Madeira, observatory in 63
magnetic compass
  see compass
magnetic deviation 17
magnetic shielding 16
Magnus/Tetens formula 26
Magowan, Cathleen 143, 144
Mainwaring, Sir Henry 88
  Sea-mans Dictionary (1644) 88
malaria-infected red blood cells 132, 142–146
Malcolm, Captain Pulteney 113
Malta 61
Manchester 214, 215, 219
Manhattan project 214
marble 21
mariner's astrolabe 228, 229, 230–231
marketing, lack of 4
Marmolada Laboratory 216, 219, 220, 223, 224, 225
Marston, Benjamin 92, 128
  "nautical dromometer" 92
Marth, Albert 61
Martin, Benjamin 92
Martin, William 130
Mason, Kenneth 201
Massachusetts Institute of Technology 214
Massey, Edward 106, 111–117, 118, 120, 124, 125, 127, 131
  bankruptcy 1821 115
  patent log 92, 95, 104, 111–117, 119,
  patent marine chronometers 116
  patent tap 115
  watchmaking patent 115
Massey, Harrie S. W. 223
materials, preservation of 23
mathematical solutions 153
mathematicians 4, 148, 150, 155, 156, 164, 167
Matsopoulos, Nikolaos 75
Mecca 159, 160, 162
  direction of 4, 148, 151, 155
  zenith of 153
mechanical logs 3, 119, 125
medical imaging 78
Mediterranean 58, 97
merchant mariners 98
Mercier, Michelle 4
meteorology 23, 25
  meteorologists 37
Meyer-Ilse, Werner 142, 143
microscope 43, 59, 187
  electron (EM) 142
  for biology, super 132, 141
  image of the IBM logo 143
  Nobel Prize 1986 143
  scanning tunnelling (STM) 142
  traditional visible light 135
  transmission x-ray (TXM) 3, 132–147
microscopy, fluorescence 146
microtechnology 137
Middle East 231
Milan 213
Miller, [T. W.], of Devonport 10
Milne, Captain [A.] 20
Milverton 190
mirror 25
Mitchelelle, Francisco of Pathhead 79
Molyneux, Samuel 97
Montague, Admiral George 114
Moon 38, 41, 65
  Eratosthenes crater 67, 68
  Maurolycus crater 65, 67
Moore, John Hamilton 85
mosques, realignment of niches 149
Morin, Paris (successor of Secretan), filar micrometer 72
movie, *A Touch of Spice*, 2003 75
Mousavi, Razieh 4
museums 1, 35, 225
  Giovanni Poleni Museum of the History of Physics, Padua 209–225
  History of Science Museum, Oxford 153, 154, 159, 160, 161, 162, 163
  Kircherian Museum, Rome 176
  Musée d'histoire des sciences, Geneva 24
  Museo Galileo, Florence 173
  National Maritime Museum, Greenwich 7, 228
  National Museum of Scotland 78, 79, 80
  Enquire gallery 78, 81
  National Technical Museum, Prague 176

Nairne, Edward 46, 47
portable equatorial 46, 48
Naples 113
Nasmyth, James 63
Nathan, Sir Matthew 203
National Observatory of Athens 2, 72–76
navy
British
see Royal Navy
Danish 13
French 13
navy crewmen, numeracy 124
Newall, Hugh Frank 2, 57, 69–71
director of the Solar Observatory 71
professor of astrophysics 71
Newall, Robert Stirling 2, 57–76
management of telescope 2, 57, 64
Newall, R. S., and Company 58
Newbiggen 58
Newcastle-under-Lyme, canal 116
Newcastle-upon-Tyne 2, 57, 60, 61
Newcomb, Simon 61
Newton, Isaac 39, 96, 119
New York 108, 109
Nicander, Henric 41, 42
Nichelson, William 87
Niemann, Bastian 140, 141
Noël, Nicolas 46
nonius 4, 167, 168
no evidence for use at sea 172
vernier scale, comparison to the nonius 177
Nordic Museum 53
exhibition "The Heritage from Newton and Linnaeus" 53
Nordmark, Zacharias 49
North
true 155
magnetic 155
North America 120
North Pole 17
Northumberland 58
North West India "frontier" 201
North-West Passage 16, 102
Norwood, Richard 85
Nova Lacerate 2010 71
Nunes, Pedro 4, 167–170
De Crepusculis (1542) 168, 169, 170
De arte atque ratione navigandih libri duo (1573) 168

octants 125
Occhialini, Guiseppe 212
Odelberg, William 53
Oestmann, Günther 228
optical tooling target 229–230, 231
oral history interviews 133
oral presentation 137
Ordnance Survey 200
Orel, Eduard von 198, 199
Orel apparatus 198
Owen, Sir [Edward] 10
Owen, William Fitzwilliam 99, 128

Padua 4, 209–225
Paris 22, 198
Paris Exposition, 1867 194
Reports of United States Commissioners 194
patents 14, 20, 21, 27, 29, 97, 100, 103, 105, 106, 108, 109, 115, 226
drawing 15
Patoun, Archibald 87
Pattinson, Hugh Lee 58
7.5-inch equatorial telescope 58
Pattinson, Mary 58
Paul, Jacques 24
Peake, E. R. L. 197
Peabody Essex Museum, Salem 110
pendulum clocks 49
Penrose, Admiral [Charles] 17
Penteli observatory 72–76
Visitor Centre 75
perambulator 118
Persian
instruction 161
scholars 155, 164
Peruzzi, Giulio 4
Philadelphia 109
Phillips, Captain [Charles] 11
Phillipson, Tacye 3
Phipps, Constantine 89, 91–92, 107, 122
photogrammetry 207
photographic camera 212
photographic equipment 193–208
photography 193
photoheliograph 194
photoresist 137, 138
photo-theodolite 194
Wild 4, 193–208
phrenological callipers 78, 80

phrenological head 78, 79
phrenology 3, 77–83
physicians 183
physicists 144, 145
physics research 3, 22
Pian di Fedaia, laboratory at 219, 220, 221, 222, 223, 224
Pimlot, J. 129
Piraeus 72
Pitot, Henri 92, 105–106
P. Kallimanopoulos Hellenic Maritime Company 72
plagiocephaly 81–82
plane-tabling 201, 206
Pointer, J. 127
Poisson, Siméon Denis 19
Poleni, Giovanni 92
pollution from factories 2
Polytechnikum of Zurich 22
Pope, William 9, 130
  dipping needle compass, 1820 13
  "Marine Automaton" 122
porous carbon 22
Porro, Ignazio 194
Portsmouth 113, 114
Portugal 167
  Alcácer do Sal 167
  Coimbra 167, 168
  Lisbon 167, 168, 174
Post Office mail boats 100, 120
  Carteret 120
  Westmoreland 120
power struggles 4
Prague 176–177
pre-Raphaelite artist 64
Preston 9
precision gears 137
pricing 3
pseudoscience 81–83
psychrometer 25, 33, 37
  Assmann aspiration 25
  August 25
  sling 25
Ptolemy's *Almagest* 168
PubMed database 81–82
Pulfrich, Carl 195
Purshull, Conyers 96, 118, 126

qibla-indicators 4, 148–166
quadrants 150, 170, 171, 172, 173–174, 176, 177
  graphical 153
Queensland Coast 203
  Cardwell 205
Qur'an 148

racial supremacy arguments 78
radioactivity 27
railway between York and London 59
Ramadan, beginning 148
  crescent Moon 148
Ramsden, Jesse 42, 46
  Ramsden circle 49
Raven, Reverend Milville 65
Rayleigh criterion 232
Red Sea 203
Reiner, Josef 24
religious rituals 148
Reneau, Lieutenant 16
Renens, Switzerland 33
Reynoldson, George 98, 126
Richard Island, Dunk Spit 205
Ringkjobing, Denmark 58
Rochefort 103
Rohrer, Heinrich 143
Roman Jesuit College 174–176
Rome 213
Röntgen, Wihelm Conrad 133
rope, four-stranded 57
Rose of Brazil, order 58
Ross, Captain James Clark 6
Ross, Captain John 16, 19, 102
Rossi, Bruno 4, 209–225
  7-ton electromagnet 210, 219, 221, 223
Rostagni, Antonio 214, 215, 216, 217, 219
Rototherm 33, 35
Rourke, E. 35
Royal Astronomical Society 58, 63
Royal College of Physicians 191
Royal Cosmographer, Portugal 167
Royal Geographical Society 193–208
  Geographical Journal 203
  Research Committee 198, 199
Royal Institute of Technology 52
Royal Institution of Great Britain 181, 190–191
  Young's eriometer 181–183

Royal Military Academy, Woolwich 7
Royal Navy 3, 7, 95, 97, 107, 114, 117
  Navy Board 16, 100, 114
  ships
  HMRY *William & Mary* 97
  HMS *Arrow* 100
  HMS *Cameleon* 100
  HMS *Donegal* 113
  HMS *Dorothea* 17
  HMS *Galatea* 113
  HMS *Hound* 99
  HMS *Isabella* 16, 100, 120
  HMS *Lord Nelson* 113
  HMS *Penelope* 104
  HMS *Princess* 113
  HMS *Racehorse* 89
  HMS *Renown* 113
  HMS *Tonnant* 16
  HMS *Trent* 17
  HMS *Vixen* 113
  HMSV *Blazer* 104
  HMSV *Lightning* 103
Royal Observatory at the Cape of Good Hope 63
Royal Society of London 58, 89, 96, 97, 179, 190, 191
  poster exhibition 53
  Philosophical Transactions of the Royal Society 46, 190
  tercentenary 53
Royal Swedish Academy of Sciences 2, 39, 40, 41, 42, 43, 52, 54
  Academy Observatory 41, 42, 44, 49
  exhibition catalogue 54
  exhibition "Scientific Links between England and Sweden in Bygone Times" 53
Rudberg, Eric 53
Rudolph, Dieter 132–138
  inventor of the TMX 136
Rudolph II, emperor 173
Rüeger, Ernst A. 32–33
Rüeger of Lausanne 21, 32
  hygrometer 33, 37
Russell, William 91–92, 100, 118, 126
Russia 22
Rymer, James 127
Sabine, Major Edward 7, 12
Safavid Iran 148–166
Safavid legal scholars 152
"Safavid world-map" 159
Salandin, Gianantonio 217
Saltsjöbaden Observatory 53, 55
sand glass, half-minute 84, 86, 87, 95
  accuracy of 86
  battery of 86
  inadequacies of 122
Saumarez, Henry de 91, 92, 95–96, 119, 120, 126
Saussure, Horace-Bénédict de 24
Savery, Thomas 96, 119
scales 174
  minutes and seconds of arc 174
Schaffer, Simon 38–39
Scheiner, Christophe 44
  helioscope 44
Schidlowsky, F. 22, 29
Schilovsky, Peter 20
Schmahl, Günter 132–138, 141, 142, 147
  inventor of the TMX 136
Schmalcalder, [John] 11
Schmidheiny, Jacob 196
Schneider, Günter 143
science 1, 77
  science publications 77
science centres 1
Scientific Instrument Commission
  2019, XXXVIII symposium, Cuba 226
  2018, XXXVII symposium, Leiden 226
scientific method 77
  scientific navigation 84
Scoresby, William 12–13
  method of producing strongly magnetised needles 12
Scots House, Gateshead 58
sea trials 107
Second World War 4, 207, 214, 216, 225
Selander, Nils Haqvin 52
semicircle, azimuthal 170
sextants 125, 170
Sharples, James 129
ship gyros, accurate alignment of 229
shipping magnates, Greek 2
Short, James 42, 46

Shuckburgh, George 46
equatorial mount 46
Siemens, Sir William 63
Sigalas, E. 72
Šíma, Zdislav 176
Sinclair, George 127
Smeaton, John 91, 120, 126
Smith, Egerton 113, 115, 116
instrument-making business 116
patent compass 116
social bias 3, 4
Société Française Stéréotopographie 198
Society of Arts 92, 100, 108, 119
Socotra, island off Yemen 204
Someda, Giovanni 212
Soret, J. L. 134
South Africa 63, 196
Drakensberg, Natal 202
South America 97
Space Shuttle 233
spark counters 27
Sparks, Captain 20
spectrograph without prisms 70
spectrometer 70
spectroscopic observations 70
of the Sun 70, 136
spectrum of Comet Halley 71
spectrum, Sun's x-ray 132
spectrum, Vega's 70
Spencer-Gregory, H. 35
Spender, M.A. 203, 206
Springer 142
Spottiswoode, William 63
Spurzheim, Johann Kaspar 77–78
St. Gall, Switzerland 27
St. Helena 107
St. John, [John Ransome] 10
Stanier, Thomas 115
Stebbing, [J. R.] Portsmouth 8, 10, 16
Steers, James Alfred 193, 202–206
diary 203, 204–206, 207
stereoscopic method 195
stereoautograph 198, 199, 201
stereo camera 204, 218
stereo-comparators 195
stereoscopic photo-theodolites 195
Stockholm 42, 44, 46
Observatory 38, 49, 52, 55
stratosphere 139
Stratton, F. J. M. 71
Strutt, J. W. 134
submarine cables, improved 58
success 1, 5, 12, 142, 197, 234
Sudan 203
sundial 159, 161, 162, 163
Sunni states 148
Sun's x-ray radiation, image 139
surveying
alpine 196
hydrographic 125
military 194
Survey of India 198
Surveyor General of India 201
surveyor 7
Svanberg, Jöns 49
Swartz, Olof 41
Sweden 38–56, 53
Administrative Court of 41
Treasury Inspection office 41, 43
Sydney, New South Wales 204, 205, 206
Syeds, [Mrs Agnes] Rotherhithe 8, 10
synchrotron storage facilities 133
Advanced Light Source, University of California, Berkeley 143

tables, mathematical 14
tachometer 206
Talas, Sofia 4
Tavaststjerna, Jacob Johan 44, 45
Tavistock 200
telescopes 3
Cassegrain 46
equatorial mounts for 44–48
Herschel telescope mount 38–56
Hubble Space Telescope 232–235
Newall telescope 57–76
glass disks for 59
new clock drive 70
observations 74, 75
pedestal for 60
three finders for 61
polar axis for 44, 45, 48, 49, 52
reflecting 2, 38–56
1.2-metre mirror telescope, Kryoneri, Corinth 75
refracting 2, 57–76

telescopes (*cont.*)
  18.5-inch refractor 61
  26-inch refractor 61
  solar telescope 138
  testing new optics 138
Téléphonie S. A. 33
Telfer, William 98, 126
temperature 31, 33
temperature range, limited 2
TESA of Lausanne 21, 32, 33
  hygrometer 33, 34, 37
theodolite 194
thermometer 24, 26, 29
  alcohol 25
  bimetallic 32
  dry bulb 25
  mercury 25
  remote telescope viewing of 26
  wet bulb 25
Thompson, Benjamin, Count Rumford 190–191
Thomson, London 10, 11
Thomson, J. J. 69
Times, The, newspaper, London 17, 120
Tirapicos, Luís 4
tools, user-friendly 150
torquetum 44, 46
transit clock 38
transit instrument 49
transversal scales 4
trigonometric problem 151
Trinity House, London 96–97
Tuscany, grand dukes of 173
Twelver Shi'i Islam 148
TXM development group 140
  Niedersachsenpries 142
Tyne, River 2, 57, 61, 64
Tyson, John 130

UNIL, the Physics Museum of the University of Lausanne 21
United Kingdom 215
United States 108, 110, 194, 214
  merchant vessels
  General Washington 109
  Merchant 110
  naval ships
  USS *Boston* 109
  USS *Philadelphia* 109
  USS *United States* 108–109
United States Naval Observatory 61
University College London 223
University of Athens, National and Kapodistrian 75
  Physics Department 75
University of Bern 21, 22
  physics professor 21
University of California, Berkeley 143, 224
  Bevatron 225
  Radiation Laboratory 224
  Luis Alvarez's bubble-chamber group 224
University of Cambridge 69–71, 73, 202, 209
  Cavendish Laboratory 27, 69, 212
  Department of Geography 202
  Emmanuel College 190
University of Coimbra 167
University of Durham 58
University of Edinburgh 223
University of Florence 210
University of Lausanne 21, 22, 27
  Physics Museum 21, 32
University of Lisbon 167
University of Padua 217
  Centre for the Study of Quick Ions 214, 216
  cloud chamber 209–225
  Institute of Physics, new building 210, 211, 212, 224
University of Salamanca 167
University of Uppsala 53
University of Zurich 27
Uraniborg 170

vacuum chamber 139
Vallet, Matthew 127
Veneto region 219
  Società Adriatica di Elettricità 219
Vetter, Johann E. 127
vernier scale 177
Violle, Jules 22
Vitruvius 92
Vlahakis, George N. 2
voltage doubler 27
Von Gedda, Baron Peter Niclas 41

Walker, Commander 11
Walker, Richard Benson 98–99, 127
Walton, Ernest 27
War Office, UK 200, 202
Washington, Captain John 95, 104
Waterloo, Battle of 125
Waters, D. W. 172
Waymouth, George 172
  "The Jewel of Artes", 1604 172
Wess, Jane 4
West Indies 113, 120
Wexler, Arnold 24, 35
Whitsunday Islands 203
Wiklund, Gustaf 38
Wilcke, Johan Carl 41, 43, 46
Wild, Heinrich 4, 196, 208
  first commercial photo-theodolite Wild P30 196
  first photogrammetric plotter. A1 stereo-autograph 196
  Wild company 196–197
  Wild T2 196–197
Wilson, Andrew 90
Wilson, Charles T. R. 209, 212, 218
wire rope machine 57
Worcester 96
Wood, [Benjamin Jasper] 11
Wood, R. W. 134
Woolley, Richard van R. 202
Woolwich Dockyard 7, 8, 9, 10, 11
Wright, Gabriel 93

x-ray 78, 132, 133, 135, 141, 142
  microscopy 142
  research units in the United States 142
  photons 133
  properties 133
  sensitive photographic film 135

Yazdī 157
  new astronomical compendium 157, 158
  quadrant scale 157
  treatise in Persian 157
York 58
Young, Thomas 4, 178–192
  damage to his reputation 191
  double-slit experiment 178
  eriometer 178–192
  applications 178
  description 183
  English medical, agricultural, textile industries, users 178–192
  learning to use 186
  physical optics 190
  principle of optical interference 178, 185
  publication on the Egyptian language 191
  status as a physician 190
  wave theory of light 4, 178, 185

Zago, Guido 216, 217, 218, 220, 222, 224, 225
Zeiss, Jena 195–196, 208
  eyepieces 72
  patents 196
  Th1 196, 197
zone-plate microscope, first 139
zone-plate telescope 132, 139
Zurich 27

Printed in the United States
by Baker & Taylor Publisher Services